Multilevel Modelling of Health Statistics

Multilevel Modelling of Health Statistics

Edited by

A.H. LEYLAND
MRC Social and Public Health Sciences Unit, University of Glasgow, UK

H. GOLDSTEIN
Institute of Education, University of London, UK

JOHN WILEY & SONS, LTD
Chichester • New York • Weinheim • Brisbane • Singapore • Toronto

This publication is designed to provide accurate and authoritative information in regard to the subject matter covered. It is sold on the understanding that the Publisher is not engaged in rendering professional services. If professional advice or other expert assistance is required, the services of a competent professional should be sought.

Other Wiley Editorial Offices

John Wiley & Sons Inc., 111 River Street, Hoboken, NJ 07030, USA

Jossey-Bass, 989 Market Street, San Francisco, CA 94103-1741, USA

Wiley-VCH Verlag GmbH, Boschstr. 12, D-69469 Weinheim, Germany

John Wiley & Sons Australia Ltd, 33 Park Road, Milton, Queensland 4064, Australia

John Wiley & Sons (Asia) Pte Ltd, 2 Clementi Loop #02-01, Jin Xing Distripark, Singapore 129809

John Wiley & Sons Canada Ltd, 22 Worcester Road, Etobicoke, Ontario, Canada M9W 1L1

British Library Cataloguing in Publication Data

A catalogue record for this book is available from the British Library

ISBN 0-471-99890-7

Typeset in 10/12pt Times by Kolan Information Services Pvt Ltd, Pondicherry, India
Printed and bound in Great Britain by TJ International Ltd, Padstow, Cornwall
This book is printed on acid-free paper responsibly manufactured from sustainable forestry in which at least two trees are planted for each one used for paper production.

Contents

Preface

Early applications of multilevel modelling tended to be concerned with educational and social rather than health data. There have been two principal motivating factors in extending the use to health. The first is that multilevel or hierarchical data structures can be found in data sets from all disciplines and the extension requires little methodological innovation. Secondly, a series of technological and methodological advances has resulted in an increase in those data that are considered to be multilevel – beyond designs featuring patients nested within hospitals and simple repeated measurements, to survival analysis and the analysis of multiple outcomes as well as hierarchies where the nesting is not strict.

The intention of this book is therefore to address the need for explanations of multilevel modelling to researchers in the health sciences interested in applying such techniques. This means that the focus is on a series of practical applications – chosen because of their relevance to such researchers – whilst sufficient explanation of the methods is also included. Although containing full references to other work, it is intended that this book should be a self-contained general reference for graduate and higher-level courses for those with a knowledge of basic regression modelling.

The book guides the reader through various stages from a basic introduction to generalised linear models, and it shows how many kinds of data can be analysed within a multilevel framework. Important statistical concepts such as the analysis of multiple (multivariate) outcomes, sampling, outliers and a review of available software are covered. Repeated measures, institutional performance, and non-hierarchical structures including spatial analysis, which also have great relevance to health and medical research, are all examined with detailed examples.

The Social and Public Health Sciences Unit is funded by the Medical Research Council and by the Chief Scientist Office of the Scottish Office Department of Health. Opinions expressed in this book are not necessarily those of either organisation. The authors are also grateful to the Economic and Social Research Council for support to the multilevel models project at the Institute of Education under the Analysis of Large and Complex Datasets programme.

Contributors

Dr William Browne
Mathematical Sciences
Institute of Education
University of London
20 Bedford Way
London WC1H 0AL
UK

Rosemary J. Day
CSERGE
School of Environmental Sciences
University of East Anglia
Norwich
NR4 7TJ
UK

Professor Harvey Goldstein
Mathematical Sciences
Institute of Education
University of London
20 Bedford Way
London WC1H 0AL
UK

Professor Michael J.R. Healy
23 Coleridge Court
Milton Road
Harpenden
Herts
AL5 5LD
UK

Professor Ita G.G. Kreft
Health and Human Services
CSLA
5151 State University Drive
Los Angeles
CA 90032–8143
USA

Dr Ian H. Langford
CSERGE
School of Environmental Sciences
University of East Anglia
Norwich
NR4 7TJ
UK

Professor Jan de Leeuw
UCLA Department of Statistics
8118 Mathematical Sciences Building
Los Angeles
CA 90095–1554
USA

Professor Toby Lewis
School of Mathematics
University of East Anglia
Norwich
NR4 7TJ
UK

Dr Alastair H. Leyland
*MRC Social and Public Health
Sciences Unit
University of Glasgow
4 Lilybank Gardens
Glasgow
G12 8RZ
UK*

Dr E. Clare Marshall
*MRC Biostatistics Unit
Institute of Public Health
University Forvie Site
Robinson Way
Cambridge CB2 2SR
UK*

Ms Alice McLeod
*MRC Social and Public Health
Sciences Unit
University of Glasgow
4 Lilybank Gardens
Glasgow
G12 8RZ
UK*

Mr John Rasbash
*Mathematical Sciences
Institute of Education
University of London
20 Bedford Way
London WC1H 0AL
UK*

Dr Nigel Rice
*Centre for Health Economics
University of York
Heslington
York
YO1 5DD
UK*

Professor Tom A.B. Snijders
*Department of Statistics and
Measurement Theory
University of Groningen
Grote Kruisstraat 2/1
9712 TS Groningen
The Netherlands*

Dr David J. Spiegelhalter
*MRC Biostatistics Unit
Institute of Public Health
University Forvie Site
Robinson Way
Cambridge CB2 2SR
UK*

Dr Geoff Woodhouse
*Mathematical Sciences
Institute of Education
University of London
20 Bedford Way
London WC1H 0AL
UK*

Dr Min Yang
*Mathematical Sciences
Institute of Education
University of London
20 Bedford Way
London WC1H 0AL
UK*

Introduction

Many data sets that occur in medical statistics or in public health and health services research display hierarchical structures; as such, it is important that these structures are taken into account using multilevel modelling. This book guides the user through the theoretical and practical aspects of multilevel modelling, and the contributors all apply the methodology to data sets relating to the health of individuals or populations.

In Chapter 1, Healy opens the volume by setting out the components of a multilevel model, explaining why such a model is useful and defining the various terms used, introducing both variance components and random coefficient models. He discusses the nature of data hierarchies that give rise to the need for such modelling and provides an example using data on children's sleep patterns.

In Chapter 2, Goldstein and Woodhouse focus specifically on the modelling of repeated measures or longitudinal data, such as observations made on the same individuals over a period of time. A basic two-level growth curve is fitted to the heights of a sample of boys, and this is then combined with a multivariate model to form a general growth prediction model including both bone age and adult height. The chapter then proceeds to consider different ways in which autocorrelation between measurement occasions can be modelled. This is followed by a discussion of various extensions to repeated measures designs.

Chapters 3 and 4 introduce generalised linear multilevel models for situations in which the responses are not continuous. In Chapter 3, Rice focuses on multilevel binomial response regression in which the response may be binary or a proportion and in the case when outcomes are observed for groups rather than individuals. Following a series of examples, a linear multilevel model for binomial data is developed and its limitations are detailed. This enables the introduction of multilevel logit and probit regression models, and, following a discussion of extra-binomial variation and model diagnostics, an example is presented concerning equity in the provision of health care services.

In Chapter 4, Langford and Day also consider discrete responses in their chapter on multilevel Poisson regression. They detail two examples: the analysis of deaths from testicular cancer among males throughout Europe in the 1970s and the weekly incidence of food poisoning in England and Wales in 1989–90.

Multilevel Modelling of Health Statistics Edited by A.H. Leyland and H. Goldstein
©2001 John Wiley & Sons, Ltd

In addition to discussing the formulation of such problems, the chapter presents substantive analyses, including the exploration of covariates and residual examination for Poisson regression models.

In Chapter 5, McLeod sets out the components of a multivariate multilevel model. She discusses how the standard multivariate regression model is generalised and shows how both continuous and discrete responses can be accommodated within the same model, together with examples relating to the blood pressure of low-birthweight children and the relationship between a patient's length of stay and probability of readmission.

In Chapter 6, Lewis and Langford set out procedures for carrying out diagnostics in multilevel models. They present methods for detecting outliers, influential data points and leverage points that affect the precision of the estimates. They explain how these quantities are calculated at each level and present examples of their use relating to mortality from prostate cancer in Europe.

In Chapter 7, Rasbash and Browne illustrate how structures that are not strictly hierarchical can be considered to be multilevel models and how such data can be modelled in this way. They discuss cross-classifications and multiple membership models, and develop general rules for the notation of such models. An example is given relating to artificial insemination.

In Chapter 8, Yang then introduces multilevel multinomial regression. An example relating to the use and abuse of antibiotics for acute respiratory tract infections illustrates a problem in which the responses are nominal, whilst ordinal responses are examined with reference to an intervention study looking at smoking prevention and cessation.

In Chapter 9, Marshall and Spiegelhalter discuss ways in which institutions (level-2 units) can be compared using multilevel models. They look at so-called 'league tables' and their statistical limitations. They show how to use information efficiently to provide comparative data that also contain valid measures of the surrounding uncertainty, demonstrating the methodology on data relating to surgical performance in New York.

In Chapter 10, Leyland shows how multilevel models can be used for the analysis of spatially organised data, and presents models for the case where there is spatial autocorrelation. He shows how such models can be expanded to include multivariate responses as well as longitudinal data. There is an example using lip cancer data from Scotland.

In Chapter 11, Snijders discusses sampling in a multilevel context. He shows how the study design is influenced by the level at which the explanatory variables are observed, and considers optimal sample sizes in multilevel models. He also considers situations where the study design is constrained by costs and when random parameters are involved, and uses an example regarding the effect of the training of psychotherapists.

In Chapter 12, Goldstein and Leyland then cover further issues in multilevel modelling not covered elsewhere within this volume. In particular, they look at meta-analysis, the modelling of survival data, and contextual and compositional effects. The chapter concludes with brief sources of references for readers interested in measurement errors, structural equation modelling and missing data in multilevel analysis.

Finally, in Chapter 13, de Leeuw and Kreft consider some of the packages, programs and modules most commonly used for fitting multilevel models. They detail the availability, documentation available, and differences between the models, interfaces and algorithms for MLwiN, HLM, VARCL, MIXFOO, MLA, BMDP5–V and PROC MIXED.

CHAPTER 1

Multilevel Data and Their Analysis

Michael J.R. Healy
Harpenden, UK

1.1 INTRODUCTION

It is a basic principle of statistics that variability in a body of data may possess
structure and that the analysis of the body of data must take account of that
structure. One of the commonest of data structures is that in which individual
units fall into a number of clusters or groups. The most familiar example in
medicine is the two-treatment clinical trial. In its simplest version, two random
samples of individual subjects (units) are drawn from the population in question and the two treatments are allocated at random to the two samples. As a
rule, it can be assumed that the variances in the two samples are equal, and the
difference between the sample means can be examined, for example by way of
an unpaired *t*-test. Alternatively, it may be possible to form pairs of subjects
(groups of two subjects each) such that the subjects in a single pair are as alike
as possible in, for example age, gender and disease severity. The analysis of the
data will now require a paired *t*-test, using the differences within each pair.
The point of this is that the pairing should have led to a decrease in the
variability within the pairs, so that to a greater extent like is being compared
with like. Because of the pairing process, the two members of a single pair are
expected to be more alike than two subjects drawn at random from the
population. Ignoring the pairing in the analysis is likely to provide an over-
estimate of the error variance and so lead to erroneous conclusions. A more
elaborate example is provided by a multicentre clinical trial. The subjects are
grouped into centres, and subjects within one centre are expected to be more
alike than subjects at different centres. There may of course be extra grouping
into pairs within the centres. In addition, it is not inconceivable that the
treatment effect differs to some extent from one centre to another, and it is
important to check for this and if necessary to take it into account as part of the
analysis.

Multilevel Modelling of Health Statistics Edited by A.H. Leyland and H. Goldstein
©2001 John Wiley & Sons, Ltd

The same kind of variability structure is common in observational as well as in experimental data. Suppose that we carry out a survey of the heights of 10-year old children, taking samples of children from the schools of a certain town. The schools constitute groups or clusters, and, for a variety of socio-economic reasons, the mean heights of different schools are likely to be different. As a result, children within a single school will be expected to be less variable in height than children drawn at random from the whole town's population of 10-year olds.

Data that are structured in this way are said to be *multilevel* or *hierarchical*. In this latter example, the hierarchy has two *levels* which we number from the bottom upwards; level 1 consists of the individual children, and level 2 consists of the schools that constitute the groups into which the children are clustered. Hierarchies may contain more than two levels. For example, in a larger survey, the schools may be grouped within towns, and these may again be grouped within larger areas.

Another example of multilevel data is provided by a situation with repeated measurements on the subjects. Suppose that we have the results of glucose determinations on patients at a number of diabetic clinics. There will be a sequence of measurements of blood glucose made on each patient. Once again, measurements made on a single patient are likely to be less variable than those made on different patients, and patients at a single clinic are likely to be more alike than patients at different clinics. Here there is a three-level hierarchy, with measurement occasions at level 1, patients at level 2 and clinics at level 3.

1.2 MODELS FOR MULTILEVEL DATA

Reverting to the two-level example of children's heights, these might be analysed by way of a simple one-way analysis of variance, distinguishing between-school and within-school variability. Such an analysis would usually be accompanied by a list of school means, each with a standard error derived from the within school mean square. An F-test could be used to test the (implausible) hypothesis that all the true school means were equal, and t-tests could be used to test for differences between schools or for more complicated contrasts. This analysis is based upon a simple mathematical model for the data, which can be written in the form

$$y_{ij} = \beta_{0j} + e_{ij}. \tag{1.1}$$

Here y_{ij} denotes the height of the ith child in the jth school, β_{0j} is a fixed quantity attached to all the children in the jth school, and e_{ij} is a random quantity attached to the ith child in the jth school. The model equation simply says that the heights of children at a particular school vary randomly about the school mean and that the means differ in an unspecified manner from school to school. The statistical model assumes that the e_{ij} are random quantities with mean zero and a variance that we shall label as σ_e^2. The quantities to be estimated in the model are the set of β_{0j} and σ_e^2.

1.3   A VARIANCE COMPONENTS MODEL

So far so good; but suppose that, rather than having studied all the schools in a single town, we have instead studied a sample of schools across a geographical region. Under these circumstances, we may not be particularly interested in the actual schools that have been included in the sample, but rather in a population of schools from which the sample has been drawn. What sort of heights might we expect to encounter, for example, if we were to measure the children in another school in the region? Now the schools are to be regarded as a random sample from a population of schools, and there are two kinds of random variability in our data: that between different children at a single school, and that between different schools. This means that we need to use a slightly different model, which we can write as

$$y_{ij} = \beta_0 + u_{0j} + e_{ij}. \tag{1.2}$$

Now β_0 is a fixed quantity applying to all the children in the study, u_{0j} is a random quantity applying to all the children in the jth school, and e_{ij} is another random quantity applying to the ith child in the jth school. The mean for the jth school, which we previously wrote as β_{0j}, is now given by $\beta_0 + u_{0j}$, indicating that it is a random member of a population of values with mean β_0. The two random quantities have zero means and they are assumed to be uncorrelated; u_{0j} has variance σ_{u0}^2 and e_{ij} has variance σ_e^2. The quantities to be estimated from the data are β_0, σ_{u0}^2 and σ_e^2. At this stage of the analysis, we are not interested in estimating quantities associated with individual schools.

This is the simplest form of a two-level model. The quantity β_0 is referred to as the *fixed part* of the model and the *u*s and *e*s constitute the *random part*. A model where the random variation is described by a set of variances is called a *variance components model* and the σ^2s are sometimes known as variance components. They are often referred to as the *random parameters* of the model.

There are several ways in which this model might be elaborated to fit more closely to reality. Although all the children in the study are between their 10th and 11th birthdays, we could take account of their exact ages by introducing another variable x_{1ij} equal to the number of days after the 10th birthday for each child. Moreover, there must be reasons for the differences between schools. If we have information on the social attributes of the neighbourhoods served by each school, we can construct an average school social deprivation score and introduce it as a third explanatory variable x_{2j} to allow for this. Notice that this variable has only a single-letter subscript. It applies to all children in the jth school and can be referred to as a level-2 variable. The elaborated model might look like this:

$$y_{ij} = (\beta_0 + \beta_1 x_{1ij} + \beta_2 x_{2j}) + (u_{0j} + e_{ij}). \tag{1.3}$$

This model has the same random part as before, but a more complicated fixed part. It takes the form of a multiple regression equation with x_{1ij} and x_{2j} as predictors, but it has in addition a level-2 error or *residual* term. In this context, note that β_0 can be regarded as the *intercept*, the expected value of y_{ij} when all the xs are equal to zero.

1.4   A RANDOM COEFFICIENT MODEL

This model can be extended further. The coefficient β_1 is the slope of the regression of height on age, and we are assuming that this is the same for each school, so that the school regression lines are parallel. It is possible that the slope, as well as the intercept, differs randomly from school to school. To allow for this, we can replace the fixed parameter β_1 by $\beta_1 + u_{1j}$, where u_{1j} is a random quantity relating to the jth school, so that the slope for the jth school is now regarded as a random quantity from a population having mean β_1 and a variance that we write as σ_{u1}^2. The whole model can now be written in the form

$$y_{ij} = (\beta_0 + \beta_1 x_{1ij} + \beta_2 x_{2j}) + (u_{0j} + u_{1j} x_{1ij} + e_{ij}), \qquad (1.4)$$

or equivalently

$$y_{ij} = \beta_{0ij} x_0 + \beta_{1j} x_{1ij} + \beta_2 x_{2ij},$$
$$\beta_{0ij} = \beta_0 + u_{0j} + e_{ij},$$
$$\beta_{1j} = \beta_1 + u_{1j},$$

with $x_0 = 1$ for all the children in the sample. The fixed part is as before, but there is now a more complicated random part that allows for the fact that both the intercepts and the slopes of the regressions on age are varying randomly from school to school. The random parameters to be estimated include the variances σ_{u0}^2, σ_{u1}^2 and σ_e^2. But we cannot assume that the random slopes and intercepts are uncorrelated, so that we also need to estimate the covariance between u_{0j} and u_{1j}, which we write as σ_{u01}. We refer to the random variables u_{0j} and u_{1j} as level-2 residuals and the e_{ij} as the level-1 residual. The u_{0j} and u_{1j} are sometimes referred to as level-2 *random effects*.

1.5   AN EXAMPLE

An example of two-level data is provided by a study kindly made available by Dr S MacKenzie (Fuller et al 1998). Children's sleep patterns were monitored on successive nights and the extent to which they coughed during the night was also recorded. Thirty-nine children were assessed on a number of nights varying from four to six. The dependent or *response* variable – the one we are aiming to explain – is the percentage of the night spent awake. As an explanatory variable we have the total number of coughs recorded during the night. Both of these variables have been transformed to logarithms (base 10) before analysis.

The data have two levels of variability: between subjects (level 2), and between nights within subjects (level 1). The study is an example of repeated measurement data. The two variables will be referred to as logawake and logcough.

A first, rather simplistic, model to consider is simply one of variance components – looking at the logawake figures by themselves: do some children tend to spend more of the night awake than others (presumably the answer is 'yes'), and if so to what extent? The model for this analysis is (1.2), where y_{ij} is the logawake variable, β_0 is the average over the population of subjects, and u_{0j} and e_{ij} are random quantities with zero means and variances σ_{u0}^2 and σ_e^2. Using the MLwiN program (Rasbash *et al.*, 1999b; see Chapter 13 for details of this and other software), the estimates of these quantities are as follows:

Parameter	Estimate	SE
β_0	0.824	0.048
σ_{u0}^2	0.068	0.020
σ_e^2	0.112	0.012

In a sample of this size, the distribution of the variance estimates is liable to be noticeably non-normal, and so the standard errors need to be used with caution. Nonetheless, the estimate of σ_{u0}^2 is more than three times its estimated standard error, and there is little doubt that the true value is greater than 0. The standard deviation within subjects is $\sqrt{0.112} = 0.33$, equivalent to more than twofold on the original scale. This is greater than the between-subject standard deviation, and illustrates the considerable variability of wakefulness from night to night within a single subject.

Let us see whether the time spent awake is affected by the amount of coughing. A possible model is now

$$y_{ij} = \beta_0 + \beta_1 x_{1ij} + u_{0j} + e_{ij}, \tag{1.5}$$

where x_1 is the logcough variable. This provides a regression line of logawake on logcough for each subject. The intercepts of the lines are random with variance σ_{u0}^2, the variance about the regression lines is σ_e^2 and the lines all have a common slope β_1. The MLwiN estimates are now as follows:

Parameter	Estimate	SE
β_0	0.671	0.059
β_1	0.138	0.034
σ_{u0}^2	0.061	0.018
σ_e^2	0.105	0.011

The slope coefficient is clearly significant – as expected, children tend to be awake more on nights on which they cough more. Both the variances have decreased – but only slightly.

We might go on to ask whether the effect of cough on wakefulness is the same for all children. To investigate this, we can allow the slope of the regression lines to vary randomly from one child to another. This means replacing the fixed coefficient β_1 by a random term of the form $\beta_1 + u_{1j}$. The residual u_{1j} varies from child to child, that is to say at level 2, and has variance σ_{u1}^2, and the model now becomes

$$y_{ij} = \beta_0 + \beta_1 x_{1ij} + u_{0j} + u_{1j} x_{1ij} + e_{ij}. \tag{1.6}$$

We now need to estimate σ_{u1}^2, and also the covariance, σ_{u10}, between u_0 and u_1. The MLwiN estimates are as follows:

Parameter	Estimate	SE
β_0	0.666	0.065
β_1	0.144	0.043
σ_{u0}^2	0.088	0.036
σ_{u1}^2	0.026	0.016
σ_{u10}	−0.027	0.020
σ_e	0.094	0.011

The interesting quantity at this stage is the estimate of σ_{u1}^2. This is greater than its standard error – but not by an impressive amount: in this sample, there appears to be not much evidence that coughing of a given amount bothers some children more than others. For what it is worth, the estimated regression lines are shown in Figure 1.1. There appear to be two subjects with negative slopes who might be investigated further. Chapter 6 looks in more detail at efficient ways of studying such 'outliers'.

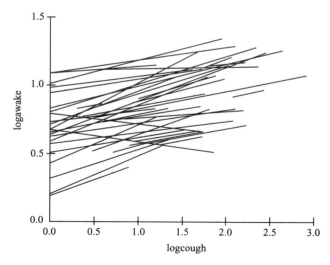

Figure 1.1 Regressions of logawake on logcough for 39 subjects.

1.6   THE LOG LIKELIHOOD

MLwiN provides a numerical value from an analysis labelled $-2*$loglikelihood. In technical language, this is minus twice the natural logarithm of the likelihood. The likelihood is, very roughly speaking, the probability of the observed data if the estimated values of the parameters in the model were the true values. This is not of particular interest in itself, but it can be very useful when it comes to comparing the fit of alternative models for the same data. Suppose we have two alternative models such that one is contained or *nested* within the other. By this is meant that the simpler model can be obtained from the more complex one simply by setting one or more of the parameter estimates to zero. Then, to a good approximation, at least in large samples, if the simpler model is true then the difference between the two values of $-2*$loglikelihood, known as the 'change in the deviance', is distributed as χ^2 with degrees of freedom given by the number of parameters that have been eliminated.

As an example we can ask with the sleep/cough data whether we need to make the regression coefficient β_1 into a random quantity at level 2 by including the random parameter σ_{u1}^2. The values of $-2*$loglikelihood are 177.1 for the model with a random coefficient and 181.7 for that in which the coefficient is fixed. The difference between these is 4.6, and if the simpler model is true then this is approximately a χ^2 with 2 degrees of freedom corresponding to the two random parameters which are included in the larger model. This gives a p-value of 0.10, not too different from that given by the comparison of σ_{u1}^2 with its standard error as shown in the previous section. Both tests are approximations but the test using the difference between log likelihoods (which is often called the *likelihood ratio test*) takes proper account of the covariance parameter and in addition is likely to provide a better approximation to the true significance probability in samples of moderate size. It can also be extended to test the effect of setting two or more parameter estimates simultaneously to zero, as with the dummy variables corresponding to the classes of a categorised predictor described below. If we wish to carry out exact inference in the case of small and moderately sized samples, we shall need to use bootstrap or MCMC methods.

1.7   RESIDUALS

Consider the ordinary (single-level) simple regression model

$$y_i = \beta_0 + \beta_1 x_i + e_i, \qquad (1.7)$$

where the βs are to be estimated and the es are random quantities with mean zero. The mean value of y for a given value of x is given by $\beta_0 + \beta_1 x$, and the difference between this and an actually observed value of y is called a *residual*. It is important to distinguish between the *true residuals* given by

$$r_i = y_i - \beta_0 - \beta_1 x_i$$

and the *estimated residuals*, which are obtained using the *estimates* $\hat{\beta}_0$ and $\hat{\beta}_1$ of β_0 and β_1 and are calculated as

$$\hat{r}_i = y_i - \hat{\beta}_0 - \hat{\beta}_1 x_i.$$

The two are not exactly equivalent; for example, the estimated residuals are restricted by the two exact relationships $\sum \hat{r}_{i=0}$ and $\sum x_i \hat{r}_i = 0$. In this context of simple regression, the residuals are mainly useful for assessing the goodness of fit of the model to the data, and a number of more or less elaborate techniques are available for examining them, many of them graphical in nature (see Chapter 6 for further discussion).

In a multilevel model, things are more complicated. To preserve as much simplicity as possible, let us revert to the two-level model

$$y_{ij} = \beta_0 + u_{0j} + e_{ij},$$

where u_{ij} and e_{ij} are random quantities that are the true level-2 and level-1 residuals with means equal to zero and variances σ_{u0}^2 and σ_e^2 respectively. In this simple model, all the y_{ij} have mean values equal to β_0, so that the *total residual* is $y_{ij} - \beta_0$ and it is estimated by $y_{ij} - \hat{\beta}_0$.

In our initial example, each of the u_{0j} was associated with a particular school. So far, we have merely considered these schools as members of a population and have used them collectively to make inferences about the population from which they were sampled. But it is entirely reasonable to be interested in one particular school, the *j*th, say. The true mean of the heights at this particular school is $\beta_0 + u_{0j}$. Neither of these two quantities is known to us, and both of them have to be estimated. The first term, β_0, constitutes the fixed part of our model, and its estimation is taken care of by whichever of the various possible model-fitting procedures we adopt. But what about the level-2 residual u_{0j}? We do not know its value, but it is possible to obtain its expected, or predicted, value given the data. This is not the same as the mean of the observed residuals at this school, the simple level-2 residual, which is given by

$$\hat{r}_j = \frac{\sum_i (y_{ij} - \hat{\beta}_0)}{n_j},$$

where n_j is the number of children sampled at this school. Instead, it turns out that, for this model, the expected level-2 residual is given by

$$\hat{u}_{0j} = \frac{n_j \sigma_{u0}^2}{n_j \sigma_{u0}^2 + \sigma_e^2} \hat{r}_j.$$

The factor multiplying $\hat{r}_j$ is less than 1, so that the estimated expected residual is less in absolute value than the simple one. This procedure is often referred to as *shrinkage*, and we speak of *shrunken residuals*.

The process of shrinkage will seem to be in need of justification. To estimate the mean height at a particular school, it seems natural to calculate the mean of the children who have been sampled at that school and to leave it at that. This is to overlook the fact that we know about the results from all other schools and that the particular school that we are interested in, insofar as it belongs to the same population, must resemble them to some extent. Examination of the shrinkage factor above shows that it becomes small (and consequently important) when σ_e^2, the within-school variability, is large compared with $n_j\sigma_{u0}^2$, which depends upon the between-school variability and on the size of the sample at the school in question. Essentially, when the sample is small and the within-school variability is large, we have relatively little information about the particular school. We do have information derived from all the other schools about the overall mean height in the population from which the school was sampled, and if we had no data at all then this mean would be the best estimate we could provide of the school's mean height. The more data we possess, the more we can trust the school's simple mean, but if the latter is not very reliable then we do well not to trust entirely the departure from the overall mean and to rely upon a compromise between the two.

Another approach may be of interest. Consider a school whose sample mean height is substantially above the overall population mean. This may occur for two reasons. Possibly the true school mean is large – its value of u_{0j} is highly positive. But all we know is the sample mean, and it is quite possible that this, by chance, is above the true school mean. If the latter is the case then the school's simple residual will be too large, and a shrunken one is to be preferred. Shrinkage will need to be appreciable precisely when the observed school mean is liable to depart appreciably from its true value – when n_j is small or when σ_{u0}^2 is large compared with σ_{u0}^2. This argument carries some conviction in the limiting case when σ_{u0}^2 is actually equal to zero and there is no true variation between schools. All the differences between the observed school means are then due to sampling variability at level 1 (within schools) and the shrunken residuals – in this case they are all zero – are clearly to be preferred.

In this simple variance components model, we have only had to consider one set of level-2 residuals. Once we complicate the model, there may be more than one of these sets of residuals, and if we wish to estimate their individual values, all of them will contain shrinkage.

In the sleep-versus-cough example we have described above, it is unlikely that we would want to focus in on a particular child. For illustrative purposes, however, we may use *MLwiN* to calculate the level-2 (subject) residuals for the simple variance component model and compare these with the simple residuals, i.e. the ordinary subject means expressed as deviations from the overall mean. The results are shown in Figure 1.2, and the shrinkage of the more extreme values towards the mean is apparent.

1.8 DUMMY VARIABLES AND INTERACTIONS

In our model for children's heights clearly it would be more realistic if we were to allow for differences in height between boys and girls. We can do this by creating an *indicator* or *dummy* variable that takes the value 0 for all the boys in the sample and 1 for all the girls. If we call this variable x_{3ij} and add a term

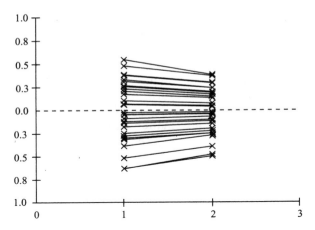

Figure 1.2 Raw (1) and shruken (2) residuals for logawake.

$\beta_3 x_{3ij}$ to the fixed part of the model then the coefficient β_3 will provide the average difference (girls − boys). In other words, β_3 represents the average height of girls relative to that of boys, taken as a *reference value*. Note that x_{3ij} is a level-1 variable, differing from one child to another in the same school, so that it has two suffices.

It may be noted that we have accounted for the two genders by using a single dummy variable. This principle can be extended. Suppose, for example, that the children were classified into five socio-economic groups labelled A, B, C, D and E. We could now create four dummy variables, x_{4ij} to x_{7ij}, say. The first of these would take the value 1 for all the children in group B and the value 0 for all the others; the second the value 1 for all the children in group C and the value 0 for all the others, and so on. Putting these variables into the fixed part, their coefficients would provide the differences between the corresponding group means and the mean of the *base* or *reference group*, in this case group A (the one for which no dummy variable has been provided). It should be noted that the choice of reference group is a matter of convenience – it does not affect the estimates of the between group differences.

Introducing the dummy variable x_3 into the fixed part of the model allows for boys and girls to have different intercepts on average. We already have a variable x_2 representing age in days after the 10th birthday, and it is quite possible in this age group that girls have different mean rates of increase of height with age than do boys. To allow for this, we need yet another variable in the fixed part of the model – one constructed as the product of the age variable x_2 and the dummy variable for gender x_3. This term in the model is called an *interaction*, specifically the interaction between gender and age.

The children in our sample have been selected as being between 10 and 11 years old, and the age variable x_2 is the number of days after their 10th birthdays; the values of x_2 thus vary between 0 and 364, and the value $x_2 = 0$ at which the intercept is measured is right at one end of the variable's range.

This produces some undesirable effects – for example, if the slope of the regression line is by chance too high, it will tend to make the intercept at $x_2 = 0$ too low, or, in technical terms, the intercept and slope coefficients will be correlated. These effects are much more marked when one of the predictors in the model is a quantity such as height or weight for which 0 is an absurd value. To avoid this, it is a good idea to measure such quantitative predictors about some convenient value near the middle of their ranges – for our age variable, we could use the number of days relative to age $10\frac{1}{2}$. This procedure is known as *centring*.

The model containing the term $\beta_3 x_{3ij}$ in the fixed part carries with it the assumption that the average gender difference between boys and girls is the same for each school. It is easy to extend the model to allow this difference to vary randomly from school to school. As might be expected, we replace the fixed coefficient β_3 by a random coefficient $\beta_{3j} = \beta_3 + u_{3j}$, where u_{3j} is a level-2 error term with mean zero, variance σ_{u3}^2 and the appropriate covariances with the other level-2 error terms.

1.9 STRUCTURES WITH COMPLEX VARIATION

Finally, consider again the model for children's heights in different schools, with age included as a predictor and with intercepts and slopes both random from school to school, i.e. at level 2. The model equation for the height of the ith child at the jth school is

$$y_{ij} = (\beta_0 + \beta_1 x_{1ij}) + (u_{0j} + u_{1j}x_{1ij} + e_{ij}).$$

It is a feature of this model that the total variability represented by the random part is not constant but depends upon an explanatory variable x_{1ij} (this variable also occurs in the fixed part of the model, but this is not an essential feature of the situation). Graphically, it can be seen that the lines with their different slopes will fan out in such a way that the variability will be greater for extreme values of the predictor than in the middle of the range. It is possible to extend this modelling of the variability structure in various different ways.

Suppose, for example, that we wish to allow boys and girls to have different amounts of level-1 variability about the regression lines. Construct the dummy variables z_{0ij} and z_{1ij}, with the first taking the value 1 for boys and 0 for girls and the second the value 0 for boys and 1 for girls. Then we can fit a model with the random part given by

$$u_{0j} + u_{1j}x_{1ij} + z_{0ij}e_{0ij} + z_{1ij}e_{1ij}, \tag{1.8}$$

where the level-1 random quantities e_{0ij} and e_{1ij} have variances σ_{e0}^2 and σ_{e1}^2 respectively and are uncorrelated. This provides different variances at level-1 for boys and girls, as we wished. There are now explanatory variables in the random part of the model that are defined at both level 1 and level 2.

There is another way in which we can represent the structure given by (1.8) that allows a flexible generalisation. Consider a model whose random part is given by

$$u_{0j} + u_{1j}x_{1ij} + e_{0ij} + e_{1ij}x_{3ij},$$

where now we assume that e_{0ij} has variance σ_{e0}^2, e_{0ij} and e_{1ij} have covariance σ_{e01}, and (somewhat bizarrely) e_{1ij} has variance equal to zero. Then the level-1 contribution to the total variance is given by

$$\sigma_{e0}^2 + 2x_{3ij}\sigma_{e01},$$

which amounts to a variance of σ_{e0}^2 for boys and one of $\sigma_{e0}^2 + 2\sigma_{e01}$ for girls (remember that the covariance may be positive or negative, so that either one of the variances may be the larger of the two). The constraining of one of the variances to equal zero while permitting a non-zero covariance is a device to introduce model complexities into the structure of the variability that can be extended in many directions. In particular, it allows us to model the level-1 variation as a linear function of several explanatory variables (for a more detailed discussion, see Goldstein, 1995, Chapter 3). The terms σ_{e0}^2 and σ_{e01} are best thought of as parameters in such a linear function rather than as variances and covariances in the usual sense.

In this chapter, the basic multilevel models have been presented; in subsequent chapters, there will be further elaborations with applications to substantive areas.

CHAPTER 2

Modelling repeated measurements

Harvey Goldstein and Geoff Woodhouse
Mathematical Sciences, Institute of Education, University of London, UK

2.1 INTRODUCTION

When measurements are repeated on the same subjects, for example students or animals, a two-level hierarchy is established with measurement repetitions or occasions as level-1 units and subjects as level-2 units. Such data are often referred to as 'longitudinal' as opposed to 'cross-sectional' where each subject is measured only once. Thus, we may have repeated measures of body weight on growing animals or children, repeated test scores on students or repeated interviews with survey respondents. Figure 2.1 is a plot of height measurements on each of four boys (Goldstein, 1989a; see below) between the ages of 10.5 and 16.5 years. Several things are worth noting. First, for each boy, the

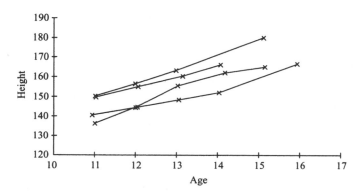

Figure 2.1   Repeated height measurements for four boys.

Multilevel Modelling of Health Statistics Edited by A.H. Leyland and H. Goldstein
©2001 John Wiley & Sons, Ltd

growth curve is very approximately a straight line. Secondly, if we drew a simple regression line for each boy, the variation about this line would be small relative to the variation between the lines. In other words the level-1 (within-individual-between-occasion) variation is smaller than the level-2 (between-individual) variation. In contrast with many other two-level data sets, where most of the variation is at level 1, we have a strong hierarchical structure where any failure to model it will result in serious model misspecification. Finally, we note that the lines for each boy have varying slopes; they grow at different rates and we will need to fit models of greater complexity than simple variance component models. Also, as we shall see below, we can fit more complex functions than straight lines to these data.

It is important to distinguish two types of model for repeated measurement data. In one, earlier measurements are treated as covariates rather than responses. In the other, as in the growth example, all the measurements are considered as responses and are related to time or age. The first case will often arise when there are a small number of distinct occasions and where *different* measures are used at each one. In this situation, it will often make little sense to study how the measures are related to age or time: to do so would require us to standardise each measurement to a common metric, but this would still leave problems of interpretation. Plewis (1993) discusses a standardisation where the coefficient of variation at each age is fixed to have a constant value. In general, however, different standardisations may be expected to lead to different inferences. The choice of standardisation is in effect a choice about the appropriate scale along which measurements can be 'equated', so any interpretation needs to recognise this.

In the second case, which is usually referred to as a 'repeated measures' model, it is more natural to ask questions about how the relationship between a common measure such as height or weight changes with age, and it is this class of models that we shall discuss here. A detailed description of the distinction between the former 'conditional' models and the latter 'unconditional' models can be found in Goldstein (1979) and Plewis (1985).

We may also have repetition at higher levels of a data hierarchy. For example, we may have annual data about smoking habits on successive cohorts of 16–year-old students in a sample of schools. In this case, the school is the level-3 unit, year is the level-2 unit and student the level-1 unit. We may even have a combination of repetitions at different levels: in the previous example, with the students themselves being questioned on successive occasions. We shall also look at an example where there are responses at both level 1 and level 2, that is specific to the occasion and to the subject.

The link with multivariate data models (see Chapter 5) is also apparent where the occasions are fixed. This can be seen in Table 2.1 where we have four measurements on each individual; the first subscript refers to occasion and the second to individual.

We can regard this as a multivariate response vector with four responses for each child, and specify a model, for example relating the measurements to a polynomial function of age. This multivariate approach has traditionally been

Table 2.1 Measurements at four occasions for three individuals.

Individual	Occasion 1	Occasion 2	Occasion 3	Occasion 4
1	y_{11}	y_{21}	y_{31}	y_{41}
2	y_{12}	y_{22}	y_{32}	y_{42}
3	y_{13}	y_{23}	y_{33}	y_{43}

used with repeated measures data (Grizzle and Allen, 1969). It cannot, however, deal with data with an arbitrary spacing of time points or number of occasions, and we shall not consider it further.

In all the models considered so far, we have assumed that the level-1 residuals are uncorrelated. For some kinds of repeated measures data, however, this assumption will not be reasonable, and we shall also investigate models that allow a serial correlation structure for these residuals.

Our examples deal only with continuous response variables, but a discussion of how to apply these procedures where responses are discrete will be given at the end of the chapter.

2.2 A TWO-LEVEL REPEATED MEASURES MODEL

Consider a data set consisting of repeated measurements of the heights of a random sample of children. Thus, for the data in Figure 2.1, we can write a simple model with linear growth as

$$y_{ij} = \beta_{0j} + \beta_{1j}x_{ij} + e_{ij}. \tag{2.1}$$

This model assumes that height Y is linearly related to age X, with each subject having their own intercept and slope, so that

$$E(\beta_{0j}) = \beta_0, \quad E(\beta_{1j}) = \beta_1,$$
$$\text{var}(\beta_{0j}) = \sigma_{u0}^2, \quad \text{var}(\beta_{1j}) = \sigma_{u1}^2, \quad \text{cov}(\beta_{0j}, \beta_{1j}) = \sigma_{u01}, \quad \text{var}(e_{ij}) = \sigma_e^2.$$

There is no restriction on the number or spacing of ages, so that we can fit a single model to subjects who may have one or several measurements. We can clearly extend (2.1) to include further explanatory variables, measured either at the occasion level, such as time of year or state of health, or at the subject level such as birthweight or gender. We can also extend the basic linear function in (2.1) to include higher-order terms and we can further model the level-1 residual so that the level-1 variance is a function of age (see Chapter 1).

Table 2.2 presents the results of an analysis fitting (2.1) and also a model that includes further polynomial growth terms. The data consist of 436 measurements of the heights of 108 boys between the ages of 11 and 16 years (Goldstein, 1989a). For convenience, age is now measured about the (approximate) mean age of 13.0 years. When we calculate polynomial terms and fit random coefficients this 'centring' will avoid numerical problems arising from approximate collinearities.

Table 2.2 Height (cm) for adolescent growth, bone age, and adult height for a sample of boys. Age measured about 13.0 years. Level-2 variances and correlations are shown. All random parameters are significant at the 5% level.

Parameter	Model A estimate (SE)			Model B estimate (SE)
Fixed				
Intercept	153.2			153.1
Age	7.10 (0.14)			7.06 (0.17)
Age^2	0.25 (0.06)			0.32 (0.06)
Age^3	−0.21 (0.02)			−0.21 (0.03)
Random				
Level 2:				
	Intercept	Age	Age^2	
Intercept	59.3			52.2
Age	0.39	0.79		
Age^2	−0.49	−0.35	0.19	
Level 1:				
σ_e^2	1.32			4.49
−2 log-like.	2182.6			2300.6

The simple 'variance components' model, which fits only an intercept at level 2, is a poor fit, as shown by the deviance statistic of 118.0 with five degrees of freedom. A variance components model, sometimes known as a 'compound symmetry' model, is anyway implausible since it assumes that the correlation between two measurements is $\sigma_{u0}^2(\sigma_{u0}^2 + \sigma_e^2)^{-1}$, the intra-unit correlation, and hence does not depend on the age difference. In fact, for these data, we can go on to fit a quartic term in the fixed part and make the coefficient of the cubic term random at the individual level; we have omitted this extended model for simplicity.

For each individual, we can estimate the posterior level-2 residuals for the intercept, linear and quadratic coefficients (see Chapter 1). Using these, we may therefore construct the predicted growth curves for each individual. Figure 2.2 shows these for the same four individuals as in Figure 2.1, and we can see the very different growth patterns.

We could go on to further elaborate this model in a straightforward way by adding covariates, for example social class, allowing us to investigate how the growth patterns vary by type of child. It is also possible to elaborate the model in an interesting way by including further response variables so defining a multivariate repeated measures model. This has a number of useful properties, as we shall explain below.

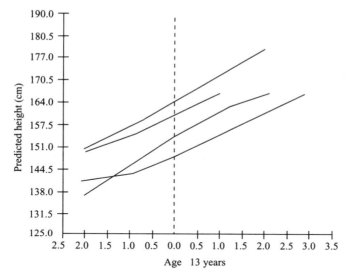

Figure 2.2 Predicted growth patterns for four individuals; fitting the model in Table 2.2.

2.3 A POLYNOMIAL MODEL EXAMPLE FOR ADOLESCENT GROWTH IN HEIGHT AND BONE AGE TOGETHER WITH ADULT HEIGHT

Our next example combines the basic two-level repeated measures model with a multivariate model to show how a general growth prediction model can be constructed. The data are as before, together with measurements of their height as adults and estimates of their bone ages at each height measurement based upon wrist radiographs. We first write down the three basic components of the model, starting with a simple repeated measures model for height using a third-degree polynomial with coefficients up to the quadratic random at the individual level:

$$y_{ij}^{(1)} = \sum_{h=0}^{3} \beta_h^{(1)} x_{ij}^h + \sum_{h=0}^{2} u_{hj}^{(1)} x_{ij}^h + e_{ij}^{(1)}, \tag{2.2}$$

where the level-1 term e_{ij} is now allowed to have a complex structure, for example a decreasing variance with increasing age, and the x_{ij}^h represent powers of the child's age.

The measure of bone age is already standardised, since the average bone age for boys of a given chronological age is equal to this age for the population. Thus we model bone age using an overall constant to detect any average departure for this group together with between-individual and within-individual variation ($u_{0j}^{(2)}$ and $u_{1j}^{(2)}$ respectively):

$$y_{ij}^{(2)} = \beta_0^{(2)} + \sum_{h=0}^{1} u_{hj}^{(2)} x_{ij}^h + e_{ij}^{(2)}. \tag{2.3}$$

For adult height, we have a simple model with an overall mean and level-2 variation given by

$$y_j^{(3)} = \beta_0^{(3)} + u_{0j}^{(3)}. \tag{2.4}$$

If we had more than one adult measurement on individuals, we would be able to estimate also the level-1 (within-individual) variation among adult height measurements; in effect measurement errors. We now combine (2.2)–(2.4) into a single model using the following indicators.

$\delta_{ij}^{(1)} = 1$ if growth period measurement,   0   otherwise;

$\delta_{ij}^{(2)} = 1$ if bone age measurement,   0   otherwise;

$\delta_j^{(3)} = 1$ if adult height measurement,   0   otherwise;

$$
\begin{aligned}
y_{ij} = {} & \delta_{ij}^{(1)} \left(\sum_{h=0}^{3} \beta_h^{(1)} + \sum_{h=0}^{2} u_{hj}^{(1)} x_{ij}^h + e_{ij}^{(1)} \right) \\
& + \delta_{ij}^{(2)} \left(\beta_0^{(2)} + \sum_{h=0}^{1} u_{hj}^{(2)} x_{ij}^h + e_{ij}^{(2)} \right) \\
& + \delta_j^{(3)} (\beta_0^{(3)} + u_{0j}^{(3)}).
\end{aligned} \tag{2.5}
$$

This is now a multivariate model, with the multivariate structure specified at level 1 using the three dummy variables above. Since adult height is defined at the individual level, its residual can only co-vary with random coefficients at that level and not at level 1. The variances and a covariance between bone age and height are specified at level 2, and the between-individual variation involving height, bone age and adult height at level 3. In fact, for simplicity, we shall assume that the residuals for bone age and height at level 2 are largely 'measurement errors', and hence it is reasonable to assume they will be independent, although dependences might arise, for example if the model were incorrectly specified at level 2. Table 2.3 shows the fixed and random parameters for this model, omitting the estimates for the between-individual variation in the quadratic and cubic coefficients of the polynomial growth curve.

From the positive value of the bone age intercept we infer that this sample is slightly advanced compared with the general population, but with a large between-individual variance of 0.70.

We see that there are non-zero correlations between adult height and both the height intercept and growth coefficients, but a smaller correlation between adult height and the bone age intercept. This suggests that the growth measurements can be used to make predictions of adult height, but that little is gained by including the bone age. To predict adult height, we require the estimated residuals for adult height from the model. For a new individual, with information

Table 2.3 Height (cm) for adolescent growth, bone age and adult height for a sample of boys. Age measured about 13.0 years. Level-2 variances and correlations shown.

Parameter	Estimate (SE)
Fixed	
Adult height:	
Intercept	174.6
Height:	
Intercept	153.1
Age	7.08 (0.16)
Age^2	0.30 (0.05)
Age^3	−0.20 (0.03)
Bone age:	
Intercept	0.21 (0.09)
Age	0.04 (0.02)

Random
Level 2:

	Adult height	Height intercept	Age	Age^2	Bone Intercept	Age
Adult height	63.4					
Height intercept	0.76	58.6				
Age	0.22	0.50	0.70			
Age^2	0.19	−0.50	−0.48	0.17		
Bone age Intercent.	0.06	0.58	0.34	−0.86	0.70	

Level 1 variances:	
Height	1.64
Bone age	0.36

available at one or more ages on height or bone age, we would estimate the adult height residual using the model parameters. This therefore provides a quite general method for predicting adult height using any collection of height and bone age measurements at a set of ages within the range fitted by the model. Table 2.4 shows the estimated standard errors associated with predictions made on the basis of varying amounts of information. It is clear that the main gain in efficiency comes with the use of height with a smaller gain from the addition of bone age.

The method can be used for any measurements, either to be predicted or as predictors. In particular, covariates such as family size or social background can be included to improve the prediction. We can also predict other events of interest, such as the estimated age at maximum growth velocity. Pan and Goldstein (1998), for example, provide estimates of growth rates and accelerations

Table 2.4 Standard errors for height predictions for
specified combinations of height and bone age measurements.

Bone age measures		Height measures (age)		
		None	11.0	11.0 12.0
None			4.3	4.2
11.0		7.9	3.9	3.8
11.0	12.0	7.9	3.7	3.7

for individuals from any set of serial measurements taken during growth. They
model height and weight in a bivariate response model and also provide
'conditional' predictions and norms for current weight or height given any set
of previous weights and heights.

2.4 MODELLING AN AUTOCORRELATION STRUCTURE AT LEVEL-1

So far we have assumed that the level-1 residuals are independent. In many
situations, however, such an assumption would be false. For growth measurements the specification of level-2 variation serves to model a separate curve for
each individual, but the between-individual variation will typically involve only
a few parameters, as in the previous example. We can think of each curve as a
smooth summary of growth with small random departures at each measurement
occasion. If, however, measurements on an individual are obtained very close
together in time, they will have a similar departure from that individual's
underlying growth curve. This implies that the level-1 residuals will be positively
correlated; there will be 'autocorrelation' between them. Examples occur in
other areas, such as economics, where measurements on each unit, for example
an enterprise or economic system, exhibit an autocorrelation structure and
where the parameters of the separate time series will vary across units at level-2.

A detailed discussion of multilevel time series models is given by Goldstein *et
al.* (1994). They discuss both the discrete-time case, where the measurements
are made at the same set of equal intervals for all level-2 units, and the continuous-time case, where the time intervals can vary. We shall develop the
continuous-time model here, since it is both more general and flexible.

To simplify the presentation, we shall drop the level-1 and -2 subscripts and
write a general model for the level-1 residuals as follows:

$$\text{cov}(e_t e_{t-s}) = \sigma_e^2 f(s). \tag{2.6}$$

This states that the covariance between two measurements s units in time apart,
depends on the level-1 variance (σ_e^2, which may be a function of age) and

Table 2.5 Some choices for the covariance function g for level-1 residuals.

$g = \beta_0 s$	For equal intervals, this is a first-order autoregressive series
$g = \beta_0 s + \beta_1(t_1 + t_2) + \beta_2(t_1^2 + t_2^2)$	For time points t_1 and t_2, this implies that the variance is a quadratic function of time
$g = \begin{cases} \beta_0 s & \text{if no replicate} \\ \beta_1 & \text{if replicate} \end{cases}$	For replicated measurements this gives an estimate of measurement reliability $\exp(-\beta_1)$
$g = (\beta_0 + \beta_1 z_{1j} + \beta_2 z_{2ij})s$	The covariance is allowed to depend on an individual level characteristic (e.g. gender) and a time-varying characteristic (e.g. season of the year or age)
$g = \begin{cases} \beta_0 s + \beta_1 s^{-1} & (s > 0) \\ 0 & (s = 0) \end{cases}$	Allows a flexible functional form, when the time intervals are not close to zero

a function involving the time difference. The latter function is conveniently described by a negative exponential reflecting the common assumption that with increasing time difference the covariance will tend to a fixed value, $\alpha \sigma_e^2$ (in the following example, we shall assume that this is zero, but in other cases this may not be reasonable):

$$f(s) = \alpha + \exp[-g(\beta, z, s)], \qquad (2.7)$$

where β is a vector of parameters for further explanatory variables z. Some choices for g are given in Table 2.5.

If we assume multivariate normality for the response variable, maximum-likelihood estimates are available (details are given by Goldstein *et al.*, 1994).

We now have a model that consists of two distinct covariance structures: the between-individual and the within-individual. From the interpretational point of view it is convenient to have parameters that summarise individual charac-teristics such as average growth and rate of growth. From this point of view, the within-individual structure exists only to provide a full description of the covar-iance structure in order to obtain a properly specified model. In some situations, however, where data may not be very extensive, we may be able to describe the overall structure *either* by fitting a small number of higher-level random coeffi-cients together with an elaborated serial correlation structure at level 1, *or* by an elaborate higher level structure and simple, independent, variation at level 1. Diggle (1988), for example, fitted a variance components model together with a level-1 serial correlation structure with $g = \beta_0 s$ to repeated measurements. Sometimes it may be possible to make a choice in terms of goodness of fit, for example using the AIC criterion based upon comparing deviances (Lindsey, 1999), but more generally the aim should be to parameterise the model so that a useful interpretation can be placed upon the parameters.

2.5 A GROWTH MODEL WITH AUTOCORRELATED RESIDUALS

The data for this example consist of a sample of 26 boys, each measured on nine occasions between the ages of 11 and 14 years (Harrison and Brush, 1990). The measurements were taken approximately three months apart. Table 2.6 shows the estimates from a model that assumes independent level-1 residuals with a constant variance. The model also includes a cosine term to model the seasonal variation in growth with time measured from the beginning of the year. If the seasonal component has amplitude α and phase γ, we can write

$$\alpha \cos(t + \gamma) = \alpha_1 \cos t - \alpha_2 \sin t.$$

In the present case, the second coefficient is estimated to be very close to zero, and is set to zero in the following model. This component results in an average growth difference between summer and winter estimated to be about 0.5 cm.

We now fit in Table 2.7 the model with $g = \beta_0 s$, which is the continuous-time version of the first-order autoregressive model.

The fixed part and level-2 estimates are little changed. The autocorrelation parameter implies that the correlation between residuals three months (0.25 years) apart is 0.18: $\exp(-\beta s) = \exp(-1.725) = 0.18$. For measurements six months apart, this drops to 0.03. This suggests that once measurements are taken less than three months apart, it will become important to fit a serial correlation model in order to specify the data structure correctly. Failure to do this will still provide consistent estimates for the fixed parameters, but will tend

Table 2.6 Height as a fourth-degree polynomial on age, measured about 13.0 years. Standard errors in parentheses; correlations in parentheses for covariance terms.

Parameter	Estimate (SE)		
Fixed			
Intercept	148.9		
Age	6.19 (0.35)		
Age^2	2.17 (0.46)		
Age^3	0.39 (0.16)		
Age^4	-1.55 (0.44)		
cos (time)	-0.24 (0.07)		
Random			
Level 2:			
	Intercept	Age	Age^2
Intercept	61.6 (17.1)		
Age	8.0 (0.61)	2.8 (0.7)	
Age^2	1.4 (0.22)	0.9 (0.67)	0.7 (0.2)
Level 1:			
σ_e^2	0.20 (0.02)		

Table 2.7 Height as a fourth-degree polynomial on age, measured about 13.0 years. Standard errors in parentheses; correlations in parentheses for covariance terms. Autocorrelation structure fitted for level-1 residuals.

Parameter	Estimate (SE)		
Fixed			
Intercept	148.9		
Age	6.19 (0.35)		
Age^2	2.16 (0.45)		
Age^3	0.39 (0.17)		
Age^4	−1.55 (0.43)		
cos (time)	−0.24 (0.07)		
Random			
Level 2:			
	Intercept	Age	Age^2
Intercept	61.5 (17.1)		
Age	7.9 (0.61)	2.7 (0.7)	
Age^2	1.5 (0.25)	0.9 (0.68)	0.6 (0.2)
Level 1:			
α_e^2	0.23 (0.04)		
β	6.90 (2.07)		

to underestimate standard errors and also not provide consistent estimates for the random parameters.

2.6   MULTIVARIATE REPEATED MEASURES MODELS

We have already discussed the bivariate repeated measures model where the level-1 residuals for the two responses are independent. In the general multivariate case where correlations at level 1 are allowed, we can fit a full multivariate model by adding a further lowest level as described in Chapter 5. For the autocorrelation model, this will involve extending the models to include cross-correlations. For example, for two response variables with the model of Table 2.7 we would write the cross-correlation as

$$g = \sigma_{e1}\sigma_{e2} \exp(-\beta_{12}s).$$

The special case of a repeated measures model where some or all occasions are fixed is of interest. We have already dealt with one example of this where adult height is treated separately from the other growth measurements. The same approach could be used with, for example, birthweight or length at birth. In some studies, all individuals may be measured at the same initial occasion, and we can choose to treat this as a covariate rather than as a response. This might be appropriate where individuals were divided into groups for different treatments following initial measurements.

2.7   CROSSOVER DESIGNS

A common procedure for comparing the effects of two different treatments, A and B, is to divide the sample of subjects randomly into two groups and then to assign A to one group followed by B, and B to the other group followed by A. The potential advantage of such a design is that the between-individual variation can be removed from the treatment comparison. A basic model for such a design with two treatments, repeated measurements on individuals and a single group effect can be written as follows:

$$y_{ij} = \beta_0 + \beta_1 x_{1ij} + \beta_2 x_{2ij} + u_{0j} + u_{2ij} x_{2ij} + e_{ij}, \tag{2.8}$$

where X_1 is a dummy variable for time period and X_2 is a dummy variable for treatment. In this model we have not modelled the responses as a function of time within treatment, but this can be added in the standard fashion described in previous sections. In the random part at level-2 we allow between-individual variation for the treatment difference, and we can also structure the level-1 variance to include autocorrelation or different variances for each treatment or time period.

One of the problems with such designs is so called 'carry-over' effects whereby exposure to an initial treatment leaves some individuals more or less likely to respond positively to the second treatment. In other words, the u_{2j} may depend on the order in which the treatments were applied. To model this, we can add an additional term to the random part of the model, say $u_{3j} \delta_{3ij}$, where δ_{3ij} is a dummy variable that is 1 when A precedes B and the second treatment is being applied, and zero otherwise. This will also have the effect of allowing level-2 variances to depend on the ordering of treatments. The extension to more than two treatment periods and more than two treatments is straightforward.

2.8   DISCRETE RESPONSE DATA

The methods we have described for continuous data can be used for discrete responses, with suitable modifications to the model and estimation procedures. We shall not go into detail here, but will sketch out a simple model for binary responses. Suppose we have data on whether or not teenage children smoke, measured at successive occasions approximately six months apart. The response is yes (1) or no (0), and we have explanatory variables such as age, gender, and social class. We may write a standard model as follows.

$$\left. \begin{array}{l} \text{logit}(\pi_{ij}) = (X\beta)_{ij} + u_j, \\ y_{ij} \sim \text{Binomial}(1, \pi_{ij}), \end{array} \right\} \tag{2.9}$$

where y_{ij} is the observed response for the jth child at the ith occasion, $(X\beta)_{ij}$ is the fixed part linear predictor containing explanatory variables, and u_j is the random effect, assumed to have a normal distribution, for the jth child measuring propensity to smoke in comparison with the population mean. The use of the logit link function is a standard procedure, and we assume that, given the fixed predictors and the individual propensity, we have independent binomial variation with probability π_{ij}, whether we observe smoking or not. Goldstein and Rasbash (1996) discuss the estimation issues for such models.

A major difficulty with (2.9) is that there will typically be many individuals who always smoke or never smoke, giving probabilities π_{ij} of 1 or 0. This implies that they have values at $\pm\infty$ for u_j. In practice, what happens if we attempt to fit such a model is that we encounter a great deal of 'underdispersion' because the level-1 variation is less than that required by the binomial assumption. One approach to this problem using a multivariate binary model is given by Yang *et al.* (2000) for the case of a small number of discrete occasions, and this approach is currently being extended to general repeated measures structures using a formulation similar to the time series model described above.

2.9 MISSING DATA

In repeated measures designs data are regarded as missing where one or more of the responses in a complete balanced design such as in Table 2.1 are unavailable. Several broad situations need to be distinguished. In the first, a response may be missing because of the study design or for reasons that are unconnected with the true, but unknown, value of the response. Thus we may deliberately design a study where each individual is measured for only a subset of occasions. Such 'rotation' designs may be practical if time is limited or where a researcher does not wish to impose too great a burden on any respondent. In other cases, the probability of being missing may depend on predictor variables in the model, but otherwise is unrelated to the model parameters, in particular the level-1 and level-2 random effects. For example, if males are more likely to have missing data than females and gender is a covariate in the model, inferences will be consistent. Situations such as these are said to involve 'ignorable' missingness.

The second situation is where the probability of a response being missing depends on the values of other observed responses. In this case, applying maximum-likelihood to the observed data yields estimates with the usual maximum-likelihood properties of consistency etc., so long as the model is properly specified.

The third situation is where the probability of an observation being missing depends on the unknown value of the observation itself. This 'non-ignorable' case is the most difficult case to deal with, and consistent estimates are possible only if one is prepared to make particular assumptions about the nature of the missingness mechanism or the distributions. Such assumptions are generally not robust, although applying a range of such assumptions as part of a general sensitivity analysis may be useful (Kenward, 1998).

In practice, care should be taken to eliminate missing data, or at least to attempt to understand its causes so that variables responsible for it can be included as covariates in a model. Little (1995) reviews the various procedures for handling missing data.

2.10   CONCLUSIONS

We have shown how very general models for repeated measures data can be constructed, including data with responses at different levels, and models where there are varying numbers of occasions and time points with the addition, where necessary, of a time series structure. We have not discussed nonlinear models such as sometimes occur in growth studies, but see Goldstein (1995) and Palmer *et al.* (1991) for a discussion of these. There are now several computer packages that will fit some or all of the models described (see Chapter 13 for a discussion).

CHAPTER 3

Binomial Regression

Nigel Rice

Centre for Health Economics, University of York, UK

3.1 INTRODUCTION

The majority of health data do not lend themselves to simple model specification allowing a linear link function to relate a set of explanatory variables to a response measured on a continuous scale. Instead, it is quite common to observe outcomes of interest that are qualitative or limited in their range of measurement. Of these, perhaps the most commonly encountered are discrete responses where an outcome may take one of a number of discrete values of either a categorical or non-categorical nature. The simplest of this type of model is one where the dependent variable is binary assuming one of two values, which, without loss of generality, may be denoted by 1 and 0, representing, for example, the presence or absence of an attribute, the success or failure of a trial, or the occurrence or not of an event.

In this chapter, we consider multilevel models in which the dependent variable assumes discrete values. For example, we may be interested in investigating the relationship between lifestyle choices, such as the intake of alcohol, smoking habits and diet, and the incidence of specific diseases such as ischaemic heart disease (IHD). In such a study, we may wish to code an occurrence of IHD as 1 and a non-occurrence as 0, and in so doing, by construction, create a binary response. The coding of such an event in this manner allows the researcher to relate a qualitative response to a set of potential explanatory variables in a regression framework and subject the resulting parameter estimates to standard statistical tests of hypotheses.

In other circumstances, instead of observing responses on individuals, we may observe the outcomes of a group of individuals or repeated experiments on the same individual. Often such data are expressed in terms of proportions, for example the proportion of patients who have shown a favourable response to a particular treatment in a clinical trial. Once again, such responses can be related to a set of explanatory variables and modelled adequately in a regression framework in much the same way as when we observe binary variables.

Multilevel Modelling of Health Statistics Edited by A.H. Leyland and H. Goldstein
©2001 John Wiley & Sons, Ltd

Underlying these methods is the assumption that the observations have a binomial distribution, and this distribution has a crucial role in the analysis of binary data.

We have seen in the preceding chapters the many advantages of correctly specifying the multilevel structure inherent in many health data. These focus on the increased substantive insight that is afforded by a fuller exploration of variability at all levels of the data hierarchy, including the explicit modelling of residual heterogeneity through the use of random coefficients. Related to this is the unbiased estimation of standard errors, which enables correct inference. Such considerations apply equally to nonlinear multilevel models such as the discrete models presented in this chapter. For example, in a Monte Carlo study using a random effects probit specification, Guilkey and Murphy (1993) showed how ignoring the random effects and performing a single-level probit analysis led to very misleading inference, since the estimated standard errors were badly biased. This is due to the single-level analysis failing to account for the correlation structure that is fundamental both in the panel data studies modelled by Guilkey and Murphy and the hierarchical data structures discussed here.

This chapter is arranged as follows. The next section presents some examples of the wide variety of circumstances where multilevel data with binary responses are encountered in the modelling of health statistics. Sections 3.3 and 3.4 consider alternative model specifications for binomial data by first discussing briefly the limitations of specifying a linear link function in the presence of a dichotomous or group dichotomous response and, secondly, presenting the alternative and commonly used logit (or logistic) and probit regression models. In Section 3.5, the central assumption of binomial variance is discussed in more detail. This includes situations where the specification of the model is thought to be correct, but where the residual deviance at level 1 does not correspond to unity, invalidating the assumption of binomial variance – a situation often referred to as extra-binomial variation. The chapter concludes with an example of measuring the extent of horizontal equity in the provision of health care services using data from the British Household Panel Survey. This is a panel survey of household behaviour collected over five years that permits a three-level analysis of repeated observations within individuals within households. The measure of health care utilisation is general practice consultations, which is dichotomised and analysed using a multilevel logit model.

3.2 EXAMPLES

With the exception of a response measured on a continuous scale, perhaps the most commonly occurring measurement in health and health care research is a discrete variable, and most frequently a dichotomous or binary variable. The analysis of such responses within a single-level regression framework has been discussed widely in the literature, and noteworthy texts on this subject include McCullagh and Nelder (1994), Cox and Snell (1989), Collet (1991) and

Maddala (1996). Developments in software for multilevel models have made the analysis of discrete response data within a hierarchical structure accessible to researchers, and selected examples of their application to health data are summarised below.

3.2.1 Geographical variations

Geographical variations in population health and health services utilisation have received considerable attention in the research literature. Much of this work is motivated by policy directives to ensure that the provision of health services correspond to criteria of equity (for example, to ensure equal access to health care services for equal health need). Gatsonis *et al.* (1993) considered interstate variation in coronary angiography utilisation for Medicare patients with a recent acute myocardial infarction. By specifying a two-level logistic model where the dependent variable represented whether individual i within state j received angiography, the authors were able to quantify state level variation and investigate the extent to which this variation was related to the age and sex of patients. Their results showed substantial interstate variation in the use of angiography after appropriate adjustment for age and sex, indicating considerable differences in the provision of medical practice across states.

3.2.2 Uptake of immunisation

In an investigation aimed at separating the ecological and individual effects in childhood immunisation uptake of pertussis (whooping cough) vaccination within a District Health Authority in England, Jones *et al.* (1991) used data on individual children clustered within treatment centres. Follow-up of children allowed outcomes to be dichotomised into those children who had received immunisation by a certain age and those who had not. By relating this binary response to a set of explanatory variables measured at both the individual and centre level through a multilevel logistic specification, an explicit investigation into the characteristics of client children, their families, and centre factors together with interaction terms helped determine which factors were important in ensuring uptake. Not surprisingly, a combination of patient, their family and centre factors were found to have played an influential role in determining uptake of immunisation, indicating that policy directives aimed at increasing the rate of uptake need to be carefully tailored to accommodate all such requirements.

3.2.3 Interpractitioner variability in prescribing

In a similar study, Davis and Gribben (1995) investigated interpractitioner variability in prescribing for given diagnoses within general practice in New Zealand. The decision to prescribe is a binary outcome that was related to characteristics of both patients and general practitioners (GPs). After controlling for these characteristics, variability in prescription rates across general practice was investigated. Their findings show that patients presenting with respiratory diseases are more likely to be given a prescription compared with patients presenting with other diagnoses, and further that variability across GP practices in the decision to prescribe is significantly greater for patients presenting with respiratory illnesses. The results of this study indicate that for at least some diagnostic categories, practice and/or GP styles have an important bearing on whether a drug is prescribed. This, the authors suggest, raises interesting questions concerning clinical uncertainty and practice autonomy, which are likely to remain central to the debate surrounding rationing in the provision of health care under scarce resources.

3.2.4 Child mortality

Studies of the determinants of infant and child mortality in developing countries have suggested that these may cluster within families. Moreover, it has been suggested that the relationship between birth spacing and infant mortality may also be affected by familial clustering of deaths. Both of these have important implications for health policy. For example, the strategic use of family planning could act to reduce not only fertility but also infant mortality by reducing the incidence of short birth intervals. Curtis *et al.* (1993) directly tested these hypotheses in an analysis of data from the 1986 Demographic and Health Survey in Brazil. For their analysis of birth spacing on infant mortality, the response variable was binary – death or survival – and observations were clustered within families, allowing a two-level logit model to be specified. The results showed a significant family random component, indicating that the probability of infant death varied across families dependent on an unobserved family effect. A comparison of the results obtained through the multilevel specification with those of a single-level logit model clearly showed that although in this example the substantive conclusions did not differ, the estimated standard errors of the single-level model were downwardly biased, reflecting the fact that the single-level model ignored important clustering within family units.

3.2.5 General practitioner incentives

Research on the effects of incentives provided to medical professionals is important in understanding how the financing and organisation of health care influences behaviour and encourages efficient medical practice. Scott and Shiell (1997a) describe an analysis of the effects a change in the remuneration system of GPs in Australia had in creating behavioural changes in the quality of consultations offered to patients. Prior to 1990, GPs were remunerated on the basis of the time it took to conduct a consultation. Since fees could be claimed for four levels of consultation, the first two of which were under 5 minutes and between 5 and 24 minutes, commentators on the health service argued that this created incentives for 'six-minute consultations'. In response to these concerns, the system was changed so that for certain participating GPs, remuneration was based on the content of the consultation rather than the time it took to perform. The expectation of this change was that incentives would be altered in such a way that more appropriate counselling and treatment of patients would ensue. To test this hypothesis, a model of the effect of fee descriptors on GP behaviour in the management of upper respiratory tract infection and sprain/strain was estimated while controlling for supply and demand-side characteristics. Three three-level logit models were considered, each consisting of repeated observations within GPs within geographical areas. The response variables chosen were whether or not a prescription was issued, whether therapeutic treatment was administered, and whether counselling was provided. GP behaviour was observed before and after the change in remuneration. Appropriate adjustments for changes in case-mix and GP characteristics were incorporated when estimating the effect of the regime change. The authors concluded that decisions made by GPs did not alter after the introduction of the change in the remuneration system and that the decisions to prescribe, treat and counsel were related to the joint characteristics of the patient and GP.

3.2.6 Competition amongst general practitioners

Scott and Shiell (1997b) adopted a multilevel probit specification to investigate the effects of competition on GPs' behaviour. The main hypothesis tested was that GPs in areas of high competition were more likely to recommend a follow-up consultation compared with GPs in areas of low competition. The lowest unit of observation was the general practice consultation which was nested within GP, which in turn was nested within geographical areas, providing a three-level hierarchical structure. Following a latent variable specification of the probit model, where y_{ijk}^* represented the difference in utility derived from recommending a follow-up visit for the ith consultation of the jth GP in the kth area, and controlling for individual, GP and area characteristics, the isolated effect of area density was sought. By stratifying by diagnostic group, the authors were able to identify the effect of competition on GP behaviour in terms of their follow-up patterns. The results lend some support to the hypothesis that GPs located in areas of high competition were more likely to recommend follow-up consultations relative to those in areas of low competition. However, this was observed only for certain medical conditions.

3.3 LINEAR MULTILEVEL MODELS

3.3.1 Binary responses

Consider the modelling of a binary response with a two-level linear specification as follows:

$$\left. \begin{array}{l} y_{ij} = \beta_{0ij} + \beta_1 x_{1ij}, \\ \beta_{0ij} = \beta_0 + u_{0j} + e_{0ij}, \end{array} \right\} \quad i = 1, \cdots, n_j, \quad j = 1, \cdots, m, \quad (3.1)$$

with $E(u_{0j}) = E(e_{0ij}) = 0$. The conditional expectation of (3.1) is given by $E(y_{ij}|x_{1ij}) = \beta_0 + \beta_1 x_{1ij}$, and we may interpret this as the probability that the event, defined by $y_{ij} = 1$, will occur, given the value of x_{1ij}. Estimated probabilities of the occurrence of the event can be computed as $\hat{y}_{ij} = \hat{\beta}_0 + \hat{\beta}_1 x_{1ij}$. However, since the dependent variable can take only one of two values (0 or 1), the residuals in (3.1) can take only one of two values: $1 - (\beta_0 + \beta_1 x_{1ij})$ and $-(\beta_0 + \beta_1 x_{1ij})$. Consequently, the variance of the error terms in (3.1) is not homoskedastic. That is, the standard assumption that $\text{var}(y_{ij}|\beta_0,\beta_1,\beta_{1ij}) = \sigma_u^2 + \sigma_e^2$ no longer holds; instead, the variance depends on the unknown probability of the event occurring. Consequently standard estimation techniques, for example IGLS (Goldstein, 1995), will not be efficient and inference may be incorrect. This, in itself, may not present too much of a problem for practical data analysis, since there are a number of ways of taking account of non-constant variance based on what are termed variance stabilising transformations (see e.g. Collett, 1991, p.51) or iteratively weighted least squares. Such methods readily extend to the multilevel model specified in (3.1).

3.3.2 Binary and proportion responses

A more important consideration in modelling binomial responses with a canonical link function is that the linear specification of (3.1) does not restrict the estimated values $\hat{y}_{ij}$ to lie within the limits [0, 1]. This is crucial if we are to interpret the conditional expectation $E(y_{ij}|x_{1ij})$ as the probability of an event occurring given the value of x_{1ij}. This will become more problematic for data with clustering near 0 or 1. The inadequacy of a linear link function to account for the truncation of data at their limits is the major drawback of specifying models of the form of (3.1) above. We see below how the specification of nonlinear functions such as the logit and probit restrict the conditional expectation to lie within the range [0, 1], and hence to be interpreted logically as estimated probabilities.

3.4 MULTILEVEL MODELS FOR BINARY AND BINOMIAL DATA

3.4.1 Binary response

We now consider a re-specification of (3.1) such that we assume an underlying latent response variable y_{ij}^* defined by the regression

$$\left.\begin{array}{l} y_{ij}^* = \beta_{0ij} + \beta_1 x_{1ij}, \\ \beta_{0ij} = \beta_0 + u_{0j} + e_{0ij}, \end{array}\right\} \quad i = 1, \ldots, n_j, \quad j = 1, \ldots, m. \quad (3.2)$$

In practice, y_{ij}^* is unobservable, and what we actually observe is a binary variable y_{ij} defined by

$$y_{ij} \begin{cases} 1 & \text{if } y_{ij}^* > 0, \\ 0 & \text{otherwise,} \end{cases} \quad (3.3)$$

such that $E(y_{ij}^*|x_{1ij}) = \beta_0 + \beta_1 x_{1ij}$. It follows from (3.2) and (3.3) that the probability of observing a 'success' defined by $y_{ij} = 1$, is

$$P(y_{ij} = 1|x_{1ij}) = P(y_{ij}^* > 0|x_{1ij}) = P(e_{0ij} > -\beta_0 - \beta_1 x_{1ij} - u_{0j})$$
$$= F(\beta_0 + \beta_1 x_{1ij} + u_{0j}) = \pi_{ij}, \qquad (3.4)$$

where $F(.)$ is the cumulative distribution function for the disturbance e_{0ij}. The observed values of y are realisations of a binomial process with probabilities given by (3.4) and vary from trial to trial (depending on x_{1ij} and u_{0j}). The corresponding likelihood function can be written as

$$L(\beta|x_{1ij}, u_{0j}) = \prod_{i=1}^{n_j} \prod_{j=1}^{m} \pi_{ij}^{y_{ij}} (1 - \pi_{ij})^{1-y_{ij}}. \qquad (3.5)$$

To progress to a marginal likelihood function, we need to impose an assumption about the distribution of u_{0j} and e_{0ij}. As in the linear case, we shall assume that the former is distributed as $N(0, \sigma_u^2)$. Then, if the cumulative distribution of e_{0ij} is assumed to be logistic, we have the multilevel logit model, and if we assume that $e_{0ij} \sim N(0, 1)$, we have the probit model.

We complete the specification of the logit model by expressing the functional form for π_{ij} in the following manner:

$$\pi_{ij} = \frac{\exp(\beta_0 + \beta_1 x_{1ij} + u_{0j})}{1 + \exp(\beta_0 + \beta_1 x_{1ij} + u_{0j})}. \qquad (3.6)$$

Correspondingly, we also have

$$1 - \pi_{ij} = \frac{1}{1 + \exp(\beta_0 + \beta_1 x_{1ij} + u_{0j})}. \qquad (3.7)$$

The observed binary responses are assumed to be binomially distributed, such that $y_{ij} \sim \text{Bin}(1, \pi_{ij})$, with conditional variance $\text{var}(y_{ij}|\pi_{ij}) = \pi_{ij}(1 - \pi_{ij})$. We can now write the multilevel logit regression model in the usual linear manner as

$$y_{ij} = \pi_{ij} + e_{0ij} Z_{0ij}, \qquad (3.8)$$

where z_{0ij} denotes the estimated binomial standard variation $z_{0ij} = \sqrt{\pi_{ij}(1 - \pi_{ij})}$. In addition, we constrain the level-1 variance to unity (i.e. we constrain $\sigma_{e0}^2 = 1$). These two terms taken together ensure the correct specification of binomial variance given above (see Goldstein, 1995).

The probit model is based upon the assumption that the disturbances e_{0ij} are independent standard normal variates, such that

$$\pi_{ij} = \phi(\beta_0 + \beta_1 x_{1ij} + u_{0j}), \qquad (3.9)$$

where $\phi(.)$ denotes the cumulative distribution function for on $N(0, 1)$ variable.

The cumulative normal distribution and the logistic distribution are very close to each other except in the tails. For both the logit and probit functions, any probability value in the range [0, 1] is transformed so that the resulting values of logit (p) and probit (p) will lie between and $-\infty$ and ∞. Both functions are symmetric about 0.5, whilst the logistic function is approximately linear in the range 0.2–0.8. In the tails of the distribution, the differences between the two functions become marked, especially for small samples where information is limited. For these reasons it is often difficult to discriminate between the two functions on the basis of goodness of fit. However, the estimates of β obtained from the two methods are not directly comparable (since the variances of the two distributions differ). To aid comparison, Amemiya (1981) suggests multiplying the estimated βs from the logit regression by 0.625, arguing that this transformation produces a close approximation between the logistic distribution and the distribution of the standard normal.

Interpretation of the parameter estimates obtained from either the logit or probit regressions are best achieved on a linear scale, such that for a logit regression, we can re-express (3.6) or (3.7) as

$$\log\left(\frac{\pi_{ij}}{1 - \pi_{ij}}\right) = \beta_0 + \beta_1 x_{1ij} + u_{0j}. \tag{3.10}$$

Equation (3.10) represents the log odds of observing the response $y_{ij} = 1$. This is linear in x_1, and so the effect of a unit change in x_{1ij} is to increase the log odds by β_1. Because the logit function is non-linear, the effect of a unit increase in x_{1ij} is harder to comprehend if measured on the probability scale given by expression (3.6) or (3.7). More simply, since the logistic function is approximately linear between 0.2 and 0.8, but nonlinear in the tails, the effect of a unit increase in x_{1ij} for π_{ij} in the linear region of the function will be different from when it is in the tails.

A further transformation of the probability scale that is sometimes useful in modelling binomial data is the complementary log–log transformation. This function again transforms a probability p in the range [0, 1] to a value in $(-\infty, \infty)$, using the relationship $\log[-\log(1 - p)]$. For small probabilities p, the complementary log–log transformation is similar to the logistic transformation, but, unlike the logistic and probit transformation, it is not symmetric about $p = 0.5$.

3.4.2 Proportions as responses

Consider the response variable representing the number of 'successes' or the number of times an event (denoted s_{ij}) occurs out of a fixed number of trials (n_{ij}), so that we have $m \times n_j$ observations of the form s_{ij}/n_{ij}. What we observe are the proportion of 'successes' or events, y_{ij} ($y_{ij} = s_{ij}/n_{ij}$). We can model the expected proportion of 'successes' (π_{ij}) using the logit link function (3.6). The observed responses are usually assumed to be binomially distributed, such that $y_{ij} \sim \mathrm{Bin}(n_{ij}, \pi_{ij})$, with conditional variance $\mathrm{var}(y_{ij}|\pi_{ij}) = \pi_{ij}(1 - \pi_{ij})/n_{ij}$. Again, we can now write the multilevel logit regression model in the usual linear manner as

$$y_{ij} = \pi_{ij} + e_{0ij} z_{0ij}, \tag{3.11}$$

where z_{0ij} denotes the estimated binomial standard deviation: $z_{0ij} = \sqrt{\pi_{ij}(1 - \pi_{ij})}/\sqrt{n_{ij}}$. Once again, we constrain the level-1 variance to unity ($\sigma_{e0}^2 = 1$) to ensure the correct specification of binomial variance. Relaxing the constraint placed on the level-1 variance – allowing it to be empirically estimated – provides us with a test of extra-binomial variation. This is the subject of the next section. The probit specification for proportions data follows in a manner analogous to the logit case.

3.5 EXTRA-BINOMIAL VARIATION AND MODEL DIAGNOSTICS

Where the assumptions of the logistic model hold, equation (3.11) provides a natural test for the assumed binomial variation. By allowing the parameter of the level-1 variance to be estimated from the data, we can compare this estimate with unity to obtain a test for extra-binomial variation. When the logistic model is thought to be correct, but the level-1 variance term is estimated to be greater than unity, the data are said to exhibit over-dispersion. Similarly, when the estimated variance is less than unity, the data are said to exhibit under-dispersion. Both are forms of variance heterogeneity. Collett (1991) provides an excellent overview of extra-binomial variation for single level logistic models.

There are a number of reasons why we may observe extra-binomial variation. Perhaps the simplest possible explanation is omitted variable bias or incorrect functional form. The former may occur if the systematic component of the model excludes variables that have strong explanatory power. These may be variables available to a study but not included in the model specification, or unobserved or unobservable variables, or relevant interaction terms that have been omitted. Alternatively, the cause of extra-binomial variation may be incorrect specification of the link function. For example, instead of a logistic link, the model may better be specified as say a complementary log–log link. Over-dispersion may also be caused by outlying data points – the detection of these in multilevel models forms the basis of Chapter 6.

Much of the intuition behind multilevel models derives from a desire to model explicitly and investigate the correlation structure that is inherent in hierarchical data. We see elsewhere in this book that failure to specify appropriately the clustering of data leads to biased estimates of variance components and standard errors. For binomial data, the same applies, and one potential cause of extra-binomial variance is through a failure to identify correctly the different levels within a model. Omitting an important level from the hierarchical structure implies that the within-group clustering of responses will not be adequately modelled and this can cause over-dispersion, since the observed number of successes can only be assumed to belong to a binomial distribution when the observations are assumed to be independent.

An important situation where extra-binomial variability in multilevel models becomes apparent is described by Wright (1997). He considers the modelling of proportions by specifying a multilevel logit function where the cell structures are sparse at level 1. In such situations, there is little information about the distributional characteristics across level-1 units within each level-2 unit. By simulating sparse and non-sparse binomial data structures with known parameters, Wright compares estimation results with true values. He concludes that sparse data structures lead to estimates of extra-binomial variation even when none is present and that, without further exploration, evidence of extra-binomial variation alone is insufficient evidence to assume model mis-specification of the sort described above. This argument is emphasised by Collett (1991), who points out that when a single-level logistic model fitted to n binomial proportions is satisfactory, the residual deviance (analogous to the residual sum of squares in linear models with a continuous response and canonical link function) has an approximate chi-squared distribution on $n - q$ degrees of freedom (q being the number of parameters in the model). Since the expected value of a chi-square random variable on $n - q$ degrees of freedom is $n - q$, it follows that for a well-specified logistic model, the mean deviance should be equal to unity. However, for sparse cell structures where proportions are based on small numbers of experimental units, the above approximation does not hold, implying that large values of the residual mean deviance may not necessarily be indicative of mis-specification. Indeed, for binary data, the deviance no longer has a chi-squared distribution, and its magnitude depends solely on the fitted probabilities.

Checking the assumption of binomial variation at level 1, where appropriate, is an important way for checking model specification. However, there are a number of other ways in which the fitted model may be inadequate. The most important of these have been mentioned above in the discussion of extra-binomial variation. In particular, the fixed-part systematic component of the model may not include variables that ought to be specified, or variables already specified that ought to be transformed. The same applies to the random part specification, in that mis-specification of the covariance structure may lead to a general mis-specification of the model. The link function chosen may not be appropriate; it may be that a logistic link function has been specified when it would have been more appropriate to use the probit specification or a complementary log–log transformation. The data may contain outlying observations or influential values that have undue impact on the fit of the model to the data (see Chapter 6). Methods for checking for each of these in single-level models are described by Collett (1991). Many of these diagnostic checks are suitable for multilevel binomial models. However, some caution is warranted. Many of the more standard methods for exploring the adequacy of the fixed-part systematic component of the model and the link function, and for detecting outliers and influential values, are in the context of modelling binomial data, where the response takes the form of proportions. Where we have binary data, many of the methods based on residuals (differences between observed and predicted values) are no longer applicable, since these do not conform, even approximately, to a normal distribution.

Two diagnostic tests that are often useful for nonlinear models of the type discussed here are the RESET and goodness-of-link test. RESET is a general test of misspecification developed by Ramsey (1969) for use in linear single-level regression models; however, it is also useful (although less informative about the type of mis-specification if present) for nonlinear models. The test statistic is calculated as follows. Predicted values from the linear index of the model under scrutiny are raised to the 2nd to 4th powers. These are then inserted as additional regressors and the model re-estimated. Under the null hypothesis that these terms are jointly equal to zero, a Wald-type test can be used to inform the suitability of the original model specification. Rejection of the null implies rejection of the model due to mis-specification error. Mis-specification may result from incorrect functional form, omitted variable bias, an inappropriate link function or inappropriate distribution of disturbance terms.

The goodness-of-link test is designed to check the adequacy of the chosen functional form assumed for the conditional disturbances at level 1 (Collett, 1991). Instead of choosing a logit function uncritically, one should consider whether a different transformation (e.g. the complementary log–log transformation) leads to a simpler model, or a model that fits the data better. For example, the appropriateness of the logit function can be appraised by computing the values

$$g_{ij} = -[1 + \hat{\pi}_{ij}^{-1} \log(1 - \hat{\pi}_{ij})]$$

and inserting g_{ij} back into the fixed part of the model and testing whether its associated parameter estimate differs significantly from zero; the null hypothesis being that it does not. Rejection of the null hypothesis implies rejection of the logit link function. The value of the associated estimated coefficient can give an indication of the likely functional form to specify.

3.6 EXAMPLE: INVESTIGATION OF EQUITY IN HEALTH CARE

Equity in the delivery of health care is a goal pursued by policy makers in most health care systems, and both how it ought to be defined and measured have received considerable attention in the research literature. However, a generally accepted view supports the notion that health care ought to be distributed according to need (degree of ill-health) and financed according to ability to pay (Wagstaff and Van Doorslaer, 1993). Empirical studies of equity have tended to focus on horizontal equity – this broadly being the principle that persons in equal need receive equal treatment, irrespective of personal characteristics irrelevant to need, such as ability to pay, race and place of residence. The investigation of evidence supporting notions of horizontal equity related to ability to pay forms the focus of the example presented here. Do persons in equal need actually receive equal treatment, or do persons with a greater ability to pay receive more than a 'fair' share of treatment? These are the questions we shall attempt to address in this example. The results given are intended for illustrative purposes only, and do not constitute a definitive analysis on this complex subject. For a more comprehensive discussion of issues surrounding equity in the delivery of health care together with an indepth international comparison of empirical evidence, the reader is referred to Van Doorslaer *et al.* (1993).

3.6.1 Data

Using five waves of the British Household Panel Survey (BHPS), the extent to which these data support the hypothesis of equity in the provision of health care services in Great Britain is addressed. The BHPS is collected by the ESRC Research Centre on Micro-social Change. 'The purpose of the BHPS is to provide a nationally representative sample of the population of Great Britain in 1991 (the year of the first wave) and to follow these individuals at yearly intervals in order to investigate processes of social and economic change' (Taylor, 1994). With information on 5500 households and over 10 000 individuals, the BHPS provides repeated observations on self-assessed and self-reported health. It also contains extensive information on household organisation, the labour market, housing, income and wealth, socio-economic values, and lifestyles. In general, the BHPS was found to be representative of the Census of 1991.

Equity is measured in relation to income as a measure of ability to pay. The measure of income used is monthly gross total family income. The BHPS provides information on the utilisation of GP services in the form of the number of visits or telephone contacts with their GP concerning their own health in the last year. Visits to hospitals are excluded. The coding of the variable is categorical, representing no visits, one or two, three to five, six to ten, more than ten visits. For the purpose of this example, the variable has been re-coded to represent 'low consulters', defined as less than three consultations per year, and 'moderate to high consulters', defined as greater than or equal to three consultations per year. Hence the outcome variable of interest is a binary variable, with 0 representing 'low consulters' and 1, 'high consulters'. The reader is referred to Chapter 8 for a discussion of the analysis of data that have been divided into more than two categories.

The BHPS section of the questionnaire on health includes an assessment of self-reported health, given in answer to the following question: 'Does your health in any way limit your daily activities compared to most people of your age?' The categorisation of this variable into 1=yes and 0=no allows a direct measure of morbidity characteristics of the sample.

Following a similar study by O'Donnell *et al.* (1993) using the General Household Survey, household income is partitioned into income quintiles. Age of individual in the study is grouped into five age ranges.

Only individuals who were sampled and present for the first-wave interview are included in the analysis. Cross-classification in the data set is ignored. Therefore individuals who changed households within the sample between two waves of data collection are treated as if they still belonged to their original household. (See Chapter 7 for a discussion of the analysis of cross-classified data in which household composition may vary between waves.) Individuals who left a household within the sample are treated as lost to follow-up. The sample data for analysis consists of 5500 households, including 10 264 individuals observed over multiple years resulting in 41 411 observations in total. Not all individuals are observed over all five waves of data collection.

3.6.2 Estimation

Estimation was performed using the software MLwiN (Goldstein et al. 1998). MLwiN estimation procedures for nonlinear models are based on Taylor series expansions to linearise the models prior to iterative generalised least-squares (IGLS) estimation (see Goldstein, 1991, 1995). All estimates were derived using second-order penalised quasi-likelihood (PQL) procedures, which are preferable to the alternative first-order marginal quasi-likelihood (MQL) methods. For a technical discussion of these issues, together with simulation results showing the improvements in moving to second-order PQL estimation, the reader is referred to Breslow and Clayton (1993), Rodriguez and Goldman, (1995) and Goldstein and Rasbash (1996). Further, where either sample sizes are small, or within-group sample sizes are small, restricted iterative generalised least-squares (RIGLS; Goldstein, 1989b) estimation is preferred to IGLS, and again this was adopted throughout. To model the relationship between the binary response and the set of explanatory variables, the logit function was used. For all models, heterogeneity in consultations across the five waves was controlled for by including the year dummies *Wave 2* to *Wave 5* in the fixed part of the model.

A limitation of the above method of estimation is that estimates of the log-likelihood for a given model specification in the presence of a binary response are unreliable. Inference is better restricted to the use of parameter estimates together with their respective estimated variances and covariances. Log-likelihood values are not reported.

3.6.3 Results

Results are presented in Table 3.1. The first column shows estimates derived from a variance components model (Model 1). From the fixed part of the model, the results clearly show the expected relationships between consultations and age and gender. In general, females consult more than males, and rates of consultation for both genders increase with age. This is reflected in the estimated coefficients of the main effects of these variables, exhibiting an increasingly positive relationship between increasing age and the probability of frequent consultations, and a positive relationship associated with the gender dummy variable *Female*. However, the interaction effects between gender and age (*Female × Age 31–45* to *Female × Age > 75*) indicate that although the probability of frequent consultations increases with age for both genders, the increase is less for females.

The self-report measure of limiting illness, *LLTI*, shows the expected strong positive association with the probability of frequent consultations. However,

Table 3.1 Three-level multilevel logistic models.

Variables	Model 1		Model 2		Model 3	
	Estimates	SE	Estimates	SE	Estimates	SE
Fixed						
Main effects:						
Constant	−2.056	0.098	−2.009	0.095	−2.022	0.096
Female	2.070	0.102	2.061	0.094	2.068	0.094
Age 31–45	0.021	0.105	−0.062	0.105	−0.057	0.105
Age 46–60	0.682	0.113	0.625	0.112	0.631	0.113
Age 61–75	1.286	0.124	1.235	0.123	1.240	0.124
Age > 75	1.807	0.176	1.754	0.180	1.760	0.182
LLTI	2.394	0.172	2.337	0.166	2.344	0.167
HHInc. – 2nd quintile	−0.158	0.058	−0.159	0.058	−0.175	0.059
HHInc. – 3rd quintile	−0.271	0.064	−0.276	0.063	−0.265	0.063
HHInc. – 4th quintile	−0.391	0.067	−0.394	0.066	−0.386	0.066
HHInc. – 5th quintile	−0.427	0.071	−0.432	0.070	−0.404	0.070
Wave 2	−0.166	0.043	−0.164	0.042	−0.165	0.042
Wave 3	−0.047	0.044	−0.047	0.044	−0.047	0.044
Wave 4	0.001	0.045	0.002	0.044	0.003	0.044
Wave 5	0.006	0.046	0.008	0.045	0.008	0.045
Interactions:						
Female × Age 31–45	−0.695	0.129	−0.622	0.125	−0.627	0.125
Female × Age 46–60	−1.203	0.141	−1.180	0.138	−1.185	0.138
Female × Age 61–75	−1.479	0.152	−1.442	0.153	−1.444	0.155
Female × Age > 75	−1.634	0.210	−1.629	0.209	−1.626	0.212
Female × LLTI	−0.349	0.123	−0.375	0.124	−0.380	0.125
LLTI × Age 31–45	0.386	0.201	0.436	0.194	0.426	0.194
LLTI × Age 46–60	0.269	0.198	0.326	0.192	0.323	0.193
LLTI × Age 61–75	−0.179	0.196	−0.128	0.192	−0.115	0.193
LLTI × Age > 75	−0.860	0.217	−0.800	0.210	−0.788	0.212
Random						
Level 1: repeated observations:						
Binomial variance	1	0	1	0	1	0
Level 2: individuals						
Variance terms:						
Constant	2.962	0.119	2.833	0.304	2.860	0.306
Covariance terms:						
Female × Constant			−0.586	0.168	−0.598	0.169
Age 31–45 × Constant			0.358	0.212	0.349	0.213
Age 46–60 × Constant			0.269	0.214	0.246	0.215
Age 61–75 × Constant			0.191	0.222	0.163	0.226
Age > 75 × Constant			0.327	0.330	0.339	0.342

Age 31–45 × Female			0.057	0.243	0.049	0.244
Age 46–60 × Female			0.565	0.262	0.584	0.263
Age 61–75 × Female			1.128	0.296	1.168	0.302
Age > 75 × Female			0.261	0.390	0.260	0.402

Level 3: Households

Variance terms:

Constant	0.869	0.101	0.760	0.095	0.849	0.185

Covariance terms:

HHInc – 2nd quintile × Constant					0.110	0.105
HHInc. – 3rd quintile × Constant					−0.050	0.105
HHInc. – 4th quintile × Constant					−0.045	0.107
HHInc. – 5th quintile × Constant					−0.148	0.107

Diagnostics

RESET, χ^2_3	4.51		5.37		5.56	
Goodness-of-link test, Z	−0.523	0.642	−0.660	0.456	−0.673	0.465
Goodness-of-link test, χ^2_1 for Z	0.663		2.100		2.100	

the effect for females, although still positive, is less than for males, as evidenced by the interaction term, *Female × LLTI*. For both genders, the effect of having a limiting illness decreases with increasing age.

Our chosen measure of equity is income; the main hypothesis being that people in equal need receive equal access to primary health care services irrespective of income. This hypothesis is tested by including household income as four dummy variables representing the quintiles of the income distribution (*HHInc–2nd quintile* to *HHInc–5th quintile*). The results show a clear income gradient with a decreasing probability of frequent consultations with a GP for increasing household income. A joint Wald test of significance of these terms gives $\chi^2_4 = 44.18$, $p = 0.0001$. The data do not support our hypothesis of no income-related inequity; instead the results suggest that inequity of access to GP services is biased towards the poor, with individuals with higher household income utilising less services for given levels of need. This result, commonly known as pro-poor inequity, has been noted elsewhere (O'Donnell *et al.* 1993).

The results of Model 1 show that the vast majority of unexplained variation in the probability of frequent consultations is between individuals within households and not between households. This result is to be expected, since the decision to visit a GP is often made on an individual basis and not determined by the household unit. However, it could be the case that for some illnesses, all household members are equally exposed, resulting in multiple members of the household consulting their GP, or that members of households share common views about attitudes towards illness or the utility of consulting their GP, both of which contribute to within-household clustering, resulting in significant between-household variation.

Model diagnostics used to check the suitability of the chosen logit specification for these data included residual plots of the level-2 and level-3 residuals as well as RESET and goodness-of-link test. A Wald-type test applied to the RESET coefficient and distributed as chi-squared on three degrees of freedom showed no evidence to suggest that the specification of the model was inadequate ($\chi_3^2 = 4.51, p = 0.21$). Similarly, a goodness-of-link test showed no evidence to suggest that the logit function was inappropriate ($\chi_3^2 = 0.663, p = 0.42$).

Of further interest in these data is the explicit modelling of the unexplained variation observed across individuals and households. For example, although the estimated probability of frequent consultations appears to increase with age, is the variability about these estimates also related to age? Similarly, we have observed that females in general have a higher estimated probability of consultations when compared to males, but is the magnitude of the level-2 variation also a function of gender? The pro-poor inequity in access to GP services observed in Model 1 tells us about the mean relationship between household income and the probability of consultations, but tells us little about how this relationship varies across households.

Model 2 presents results from including level-2 random coefficients to represent age and gender effects. We observe an overall level-2 variance constant term together with covariance terms representing the departure in variance for females *(Female × Constant)*, age group *(Age 31–45 × Constant* to *Age > 75× Constant)* and female by age interaction terms *(Age 31–45 × Female* to *Age > 75 × Female)*. We can test the significance of these terms using Wald-type tests, distributed as chi-squared statistics. Correspondingly, a test of significance of the age terms resulted in $\chi_4^2 = 3.28$, which is not significant ($p = 0.51$). However, the age by female covariance terms are highly significant, with $\chi_4^2 = 17.8, p = 0.001$. The female main effect is also highly significant: $\chi_1^2 = 12.14, p = 0.0005$. The negative coefficient on the female dummy variable indicates that although, in general, females had a higher probability of frequent consultations, they showed less variability in their pattern of consultations than males. This result, however, neglects the age effects and interaction terms between age and gender. Younger females show less variability in their consultation patterns than males, but this increases with age until the age group 61–75, in which variability about female consultation probabilities is significantly greater than for males.

The results of Model 2 taken on the whole show that although females have a higher probability of frequent consultations at all ages, the variation across individuals is far greater for males than females with the exception of the 61–75 age group.

Model diagnostics show little evidence to suggest that the specification is incorrect. Some indication, although very small, of the gain in efficiency of the fixed part parameter estimates from including random coefficients at level 2 can be seen from comparing the standard errors of the fixed-part estimators in Model 2 with Model 1.

In Model 3 we observe the effect of specifying the household income quintiles as random components at level 3. A joint test of significance of these terms gives $\chi_4^2 = 7.00, p = 0.14$. The null hypothesis that unexplained variation across households is not a function of household income therefore cannot be rejected. Despite evidence suggesting there exists pro-poor inequality in access to GP services, there is no evidence to suggest that variance in access observed across households is also related to income. Model 3 appears to be well specified, but, owing to the random coefficients at level 3 contributing very little, Model 2 is to be preferred.

3.6.4   Conclusions

The results of this analysis suggest that there exist pro-poor inequalities in access to primary care services. However, these results warrant some caution. Standardisation for health care need is achieved through the use of self-reported limiting illness. This is one important aspect of health care need, but is far from being the only one. To draw firm conclusions from any analysis of this sort, better and more encompassing measures of health care need are required. Further, the use of gross household income is justified on the basis that this, rather than individual income, determines consumption patterns that affect all household members. Other studies of health inequalities have questioned the use of such measures, arguing that permanent income, rather than instantaneous income, or life-time consumption possibilities are more relevant measures (Steen Carlsson and Lytkens, 1997; Modigliani and Ando 1963).

Further, the estimation procedure used in this example is generally robust where sample sizes are large. For small sample sizes or where there are few lower-level units within each higher-level unit and for binary responses, the Taylor series approximation may produce downwardly biased parameter estimates. Recent work on further improving the estimates from quasi-likelihood procedures using an iterative bias-corrected bootstrap have proved extremely effective in removing this bias (Goldstein *et al.*, 1998). However, these methods are computationally demanding, and at present are only feasible for small to medium-sized data sets. Other approaches, such as Bayesian Markov chain Monte Carlo (MCMC) methods using Gibbs and Metropolis-Hastings sampling, are also useful for obtaining unbiased estimates (see e.g. Gilks *et al.* 1996). These methods, although again computationally intensive, offer alternative procedures for binomial data and, in general, it is recommended that more than one approach is tried; if similar results are obtained then more confidence can be placed in the estimates.

CHAPTER 4

Poisson Regression

Ian H. Langford and Rosemary J. Day
CSERGE, School of Environmental Sciences, University of East Anglia, UK

4.1 INTRODUCTION

There are many instances in health research where the data consist of counts of
a particular disease or state of health. For example, we may have the number of
individuals who develop cancer within a particular area for a given time period.
We may then wish to compare this with the numbers of people developing
cancer in other areas, perhaps over the same or a similar time period. If each
area is very small, and contains only a few individuals, we could consider the
data to be binomially distributed. However, as the number of persons at risk in
each area increases, perhaps to hundreds, thousands or tens of thousands, it is
sensible to make use of the Poisson distribution to model the counts or cases of
a disease occurring in that population, particularly if the disease is rare. We
define 'rare' as meaning that some of the areas we are interested in have less
than 10 cases occurring within the time period of our study (Clayton and Hills,
1993).

For any area, we can define the probability that x events occur within the
time period of our study as being

$$P_x = \frac{\mu^x e^{-\mu}}{x!}.$$ (4.1)

The Poisson distribution has some properties that we can make use of when
modelling our data. For example, the expected or mean value of x is equal to
the variance of x, so that

$$E(x) = \text{var}(x) = \mu.$$

In the first example that we consider in this chapter, we have the number of
deaths of males from testis cancer in a number of European countries during
the 1970s (Smans *et al.*, 1992) occurring in the male population at risk in level
III areas, as defined by EEC statistical services. These are the equivalent of
counties in England and Wales, départments in France and Regierungsbezirk

Multilevel Modelling of Health Statistics Edited by A.H. Leyland and H. Goldstein
©2001 John Wiley & Sons, Ltd

in (West) Germany. A simple way of modelling the data would be to compare the mortality rates from testis cancer across all the areas included in our study. However, this would ignore an important feature of the data, namely that the areas are grouped into an administrative hierarchy, where level III units are nested within level II units (regions), and these in turn are nested within nations. This geographical structure to the data is important for a number of reasons, including the following:

(a) There may be different diagnostic procedures or treatment regimes in different administrative areas, which may be regions or nations, and hence there may be different survival or recording rates between regions or nations.
(b) There may be causal factors that operate at regional or national scales, for example, differences in lifestyle, diet, exposure to environmental pathogens or genetic composition, which may explain some of the variation seen in our data.

If we ignore the possibility that processes may be operating at these larger geographical scales then we may draw incorrect inferences from an analysis of data only at county level. Chapter 10 examines how we may generally consider spatially correlated effects, but here we consider only hierarchical effects due to the inherently multilevel structure of our data.

Some data may occur as counts where we do not have to take into account a denominator population. In our second example, we examine the weekly incidence of food poisoning for 401 districts in England and Wales (OPCS, 1989, 1990). In this case, we can assume that the denominator population of each district remained roughly constant over time. However, we still have a multilevel model, because weekly reports of the incidence of food poisoning are nested within districts, i.e. we have a time series for each district, and we wish to see if there is any relationship between food poisoning and mean weekly temperature that remains constant across districts, and also how the relationship varies between districts.

In the following section, we develop the models required to examine Poisson distributed data in the multilevel case. Then we provide detailed analyses of the two models described above. The discussion focuses on methodological rather than substantive issues, and details some recent developments and pointers for future research. All the models fitted in this chapter were done so using the MLwiN software (Goldstein *et al.*, 1998).

4.2 METHODS

When we have Poisson distributed data, it is usual to use a logarithmic transformation to model the mean, i.e. $\log \mu$. This is the natural parameter for modelling the Poisson distribution (Dobson, 1991; McCullagh and Nelder, 1994), but there are other good reasons for using a logarithmic transformation.

Suppose we have a set of areas, indexed i, with O_i cases in the ith district, and E_i expected cases in the denominator population, where E_i may be standardised by characteristics such as age and sex of the population, and $\sum_i O_i = \sum_i E_i$. The relative risk for the disease of interest is then

$$\theta_i = \frac{O_i}{E_i}. \tag{4.2}$$

If we use a logarithmic transformation then we prevent our model from predicting negative numbers of cases of the disease for any area, and hence negative relative risks, which cannot exist. A logarithmic transformation also means that any effects on the distribution of the disease that we include as explanatory variables in the model are multiplicative – an assumption that is often in accord with epidemiological theory. However, there is no theoretical restriction on using other transformations of μ, as discussed in Dobson (1991). In this chapter, however, we shall use a logarithmic transformation, so that we can write down a simple single-level Poisson model as:

$$\left.\begin{aligned}
O_i &\sim \text{Poisson}(\pi_i), \\
O_i &= \pi_i + e_{0i}x_{0i}^*, \\
\log \pi_i &= \log E_i + \beta_0 + \beta_1 x_{1i}, \\
x_{0i}^* &= \pi_i^{0.5},
\end{aligned}\right\} \tag{4.3}$$

where β_0 is an intercept parameter, and β_1 is a slope parameter associated with a continuous variable x_1. The term $\log E_i$ is included in the model as an offset, with a parameter fixed as 1 to account for the different populations at risk of the disease in each area. The variable x_0^* is a vector comprising the square root of the predicted values π_i to allow for Poisson distribution of the cases O_i, i.e. $\text{var}(O_i) = \pi_i$. In addition, if we assume there is only Poisson variation in the data, the variance of the residuals, $\text{var}(e_{0i}) = 1$. We can extend this to the two-level case by assuming we have units at a lower level, indexed by i (e.g. counties) nested within higher-level units indexed by j (e.g. regions). We can write an equivalent model as

$$\left.\begin{aligned}
O_{ij} &\sim \text{Poisson}(\pi_{ij}), \\
O_{ij} &= \pi_{ij} + e_{0ij}x_{0ij}^*, \\
\log \pi_{ij} &= \log E_{ij} + \beta_{0j} + \beta_{1j}x_{1ij}, \\
\beta_{0j} &= \beta_0 + u_{0j}, \\
\beta_{1j} &= \beta_1 + u_{1j}, \\
\begin{bmatrix} u_{0j} \\ u_{1j} \end{bmatrix} &\sim N(0, \Omega_u), \quad \Omega_u = \begin{bmatrix} \sigma_{u0}^2 & \sigma_{u01} \\ \sigma_{u01} & \sigma_{u1}^2 \end{bmatrix}, \\
x_{0ij}^* &= \pi_{ij}^{0.5},
\end{aligned}\right\} \tag{4.4}$$

which we estimate using an iterative generalised least-squares procedure (Goldstein, 1995). In this case, we model the Poisson variation at level 1, and assume the variation at higher levels to be multivariate normally distributed. There are several interesting features of the model. First, we can have random intercepts and slopes at the higher level. This means that we estimate fixed effects as in a single-level model, so we have a mean intercept β_0 and a mean slope β_1 around which we measure residual intercepts u_{0j} and residual slopes u_{1j} for each of the higher-level units. Hence we are not constrained to fitting one relationship for the whole data set, but can estimate separate relationships for each area, if that is the purpose of our study. We can see if there is significant overall variation of the intercepts and slopes for the higher-level units by examining the parameters σ_{u0}^2 and σ_{u1}^2. We can also estimate the covariance between the intercepts and slopes, σ_{u01}.

However, although we have modelled complex variation at level 2 in our model, we have so far assumed that there is only Poisson variation at level 1. Extra-Poisson variation may occur for a number of reasons, one being that we have areas with very different sizes of population at risk, i.e. the E_i are heterogeneous (Clayton and Kaldor, 1987; Langford 1994). In this case, we measure a variance parameter at level 1 of our model, so that

$$\left. \begin{aligned} O_{ij} &= \pi_{ij} + e_{0ij}\pi_{ij}^{0.5}, \\ \text{var}(e_{0ij}) &= \sigma_{e0}^2. \end{aligned} \right\} \tag{4.5}$$

This means that we are estimating a constant to multiply by the Poisson variance we measure at level 1. However, there may be occasions (see e.g. Langford and Bentham, 1997) where this is not sufficient to account for all the variation seen in the data, particularly if there are a few areas with very large population amongst a sample of mainly small areas, as occurs with data for local authority districts in the UK census (see the example in Section 4.3.2). In this case, we can estimate negative binomial variation at level 1, where we model

$$\left. \begin{aligned} O_{ij} &= \pi_{ij} + e_{0ij}\pi_{ij}^{0.5} + e_{2ij}\pi_{ij}, \\ \text{var}(e_{0ij}) &= \sigma_{e0}^2, \\ \text{var}(e_{2ij}) &= \sigma_{e2}^2, \end{aligned} \right\} \tag{4.6}$$

and we have a quadratic expression for the variance at level 1.

One problem remains, and that concerns the estimation of residuals at higher levels of the model, since these are nonlinear with respect to the responses. This is solved by using a Taylor series approximation to estimate the level-2 residuals (Goldstein 1995; Goldstein and Rasbash, 1996; Langford et al., 1998). The technical discussion of this issue is beyond the scope of this chapter, but we detail here the choices that are available in the MLwiN software; these are as follows:

(a)   The Taylor series approximation can be first- or second-order, where the second-order approximation includes an extra term in the expansion, and is hence more accurate, but increased instability means that it is less likely that the model will converge.

(b) The estimation procedure can be done using marginal quasi-likelihood (MQL), where estimation of the residuals is based around the fixed part of the model only, or penalised quasi-likelihood (PQL), where estimation is based around the fixed part of the model and the current estimates of the higher-level residuals. Second-order PQL estimation is the most accurate approximation, but the one most likely to incur problems of convergence, particularly if there are one or more large residuals estimated in the model.

4.3   EXAMPLES

4.3.1   Testis cancer mortality in the European Community

Data on testis cancer mortality for males of all ages were taken from the *Atlas of Cancer Mortality in the European Community* (Smans *et al.*, 1992). Testis cancer mortalities were defined as being deaths recorded and certified by a medical practitioner and coded as ICD8-186. These data were collected between 1971 and 1980, although for the UK, Ireland, Germany, Italy and The Netherlands, data were only available from 1975–1976 onwards, and are aggregated over the period of data collection. The geographical resolution of the data are level III EC units, roughly represented by the county in the UK (indexed *i*), which represent the lowest level (level 1) of our multilevel model. These are nested within regions (level 2 of our model, indexed *j*), which in turn are nested within nations (level 3, indexed *k*). The multilevel hierarchy of the data is illustrated in Table 4.1.

We were interested in the distribution of testis cancer mortality in relation to income and urban–rural status. Therefore we included two explanatory variables in our model that are measured in the nine EC countries included in our analysis. These are *RGDP*, which represents GDP per inhabitant, and *RDENS*, the density of inhabitants per square kilometre. These are both measured at region level, level 2 of our model. The model we fitted to the data is a variance components model, as follows:

Table 4.1 The geographical hierarchy of the EC mortality data.

Nation	Regions	Counties
1. Belgium	3	11
2. West Germany	11	30
3. Denmark	3	14
4. France	21	94
5. United Kingdom	11	70
6. Italy	20	95
7. Ireland	4	26
8. Luxembourg	1	3
9. Netherlands	4	11

$$
\left.\begin{aligned}
O_{ijk} &\sim \text{Poisson}(\pi_{ijk}), \\
O_{ijk} &= \pi_{ijk} + e_{0ijk}\pi_{ijk}^{0.5}, \\
\log \beta_{ijk} &= \log E_{ijk} + \beta_{0jk} + \beta_1\, RGDP_{jk} + \beta_2\, RDENS_{jk}, \\
\beta_{0jk} &= \beta_0 + v_{0k} + u_{0jk}, \\
v_{0k} &\sim N(0, \sigma_v^2), \\
u_{0jk} &\sim N(0, \sigma_u^2).
\end{aligned}\right\} \qquad (4.7)
$$

In Table 4.2, we present the results from fitting the model using the different estimation procedures described. The first model in Table 4.2 is fitted using the MQL procedure with a first-order Taylor series approximation, and the variance has been constrained to be Poisson. The term β_0 refers to the mean intercept for the data. β_1 is the mean, or fixed slope for the explanatory variable $RGDP$, and β_2 is the mean slope for the variable $RDENS$. β_1 is positive and significant at the 0.05 level, whilst β_2 is negative and significant at 0.10 level. We can see that there seems to be variance occurring at both regional and national levels, with about four times as much variance being seen between nations as between regions. However, we have many more regions than nations, and the variance at regional level is statistically significant at the 0.05 level, whilst, from a simple Wald test, dividing the parameter estimate by its standard error, the variance between nations is only significant at the 0.10 level. However, it is worth retaining this parameter, since we have only nine nations present at level 3, and the Wald test may not be reliable in assessing the significance of this parameter.

When we allow for extra Poisson variation (EPV), we see that a level-1 variance parameter of 1.36 is estimated, which is more than 1.96 standard deviations away from its expected value of 1, indicating that there is significant EPV between level-1 counties. We also note that the variance at regional level has dropped slightly when this EPV is included in the model. The fixed parameters in the model have only been slightly changed. Using a more precise second-order approximation, we note that the EPV is estimated at a higher value of 1.45. Using a first-order PQL approximation, we see that the variance estimates are very similar to the MQL^2 approximation, but the fixed parameters have both deviated further from zero, whilst their standard errors have increased. The best approximation, second-order PQL, shows similar values for the fixed parameters, whilst the variance at levels 1 and 2 is slightly higher than for the first-order MQL, for example.

We can also estimate residuals from the model, and place confidence limits around them (for an explanation of this procedure, see Goldstein, 1995). Figure 4.1 shows the residuals at level 2 for the intercept term plotted against their rank order, with 95% confidence bounds. As we can see, there are only a few regions that are significantly high or low compared with the mean (zero on the graph), at the far right and far left of the graph respectively. We can also extract level-3 residuals for individual nations, and these are shown in Figure 4.2. Denmark and West Germany appear to be significantly higher than the mean, whilst Italy and particularly Belgium at the extreme left are lower than the overall mean.

Table 4.2 Variance components model for mortality from testis cancer using different estimation methods.

	MQL^1 Poisson		MQL^1 EPV		MQL^2 EPV		PQL^1 EPV		PQL^2 EPV	
	Estimate	SE	Estimate	SE	Estimate	SE	Estimate	SE	Estimate	SE
Fixed part										
β_0	2.65	0.11	2.66	0.11	2.59	0.12	2.60	0.11	2.58	0.11
$\beta_1 (\times 10^{-3})$	3.35	1.27	3.24	1.29	3.24	1.37	3.59	1.41	3.61	1.42
$\beta_2 (\times 10^{-5})$	-6.86	3.94	-6.77	4.12	-6.77	4.40	-7.11	4.66	-7.22	4.71
Random part										
Level 3: nations										
σ_v^2	0.098	0.052	0.101	0.054	0.109	0.061	0.096	0.052	0.096	0.052
Level 2: regions										
σ_u^2	0.027	0.007	0.024	0.007	0.027	0.008	0.028	0.008	0.028	0.008
Level 1: counties										
σ_e^2	1	0	1.36	0.11	1.45	0.12	1.45	0.12	1.48	0.12

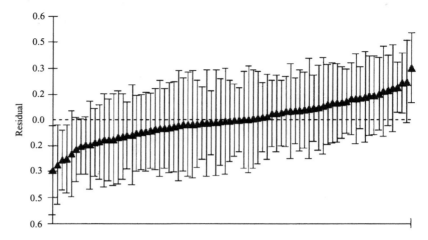

Figure 4.1 Residuals and 95% confidence intervals for the 78 regions.

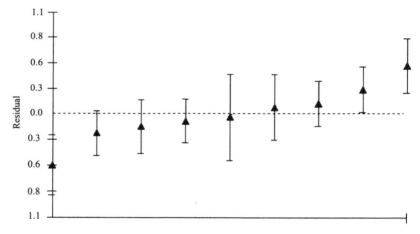

Figure 4.2 Residuals and 95% confidence intervals for the 9 countries.

We can also fit random coefficients for the explanatory variables into the model at regional and national levels. When this is done, we find that no parameters are estimated at regional level, but there is a variance and covariance estimated at national level for *RGDP*, as shown in Table 4.3. As can be seen from Table 4.3, these random effects are not large, but are interesting for demonstration purposes.

From the model in Table 4.3, we can see that the fixed parameter for *RGDP* has decreased in value, but that there are now separate intercepts and slopes for different nations. These slopes and intercepts are shown graphically in Figure 4.3. Note that *RGDP* has been centred around its mean value, which aids convergence of the model (Goldstein, 1995). The slopes for each country look fairly similar, except for Italy, which has a steeper slope, and hence a more

Table 4.3 Random coefficients model for mortality from testis cancer.

	PQL^2 EPV	
	Estimate	SE
Fixed part		
β_0	2.56	0.12
$\beta_1(\times 10^{-3})$	2.65	1.91
$\beta_2(\times 10^{-5})$	−4.74	4.92
Random part		
Level 3: nations		
σ_{v0}^2	0.104	0.057
$\sigma_{v01}(\times 10^{-4})$	−1.10	5.82
$\sigma_{v1}^2(\times 10^{-6})$	8.40	10.89
Level 2: regions		
σ_u^2	0.024	0.008
Level 1: counties		
σ_e^2	1.49	0.12

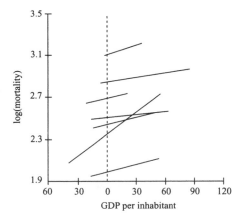

Figure 4.3 Relationship between GDP (centred) and mortality.

positive relationship between testis cancer mortality and regional *RGDP*. This slope is significantly different from the mean slope for all nations, and one possibility for further modelling would be to include a separate intercept and slope for Italy, which will not be followed up here (but see Chapter 6 for a discussion of outliers).

4.3.2 Weekly incidence of food poisoning in England and Wales

Data on the reported weekly number of cases of food poisoning for 54 administrative counties of England and Wales for the period 1989–90 were taken from published sources (OPCS, 1989, 1990). These years were chosen because ambient temperatures, especially in the summer months, were unusually high, and we wished to examine whether there is a relationship between incidence of food poisoning and ambient air temperatures. Previous work by Bentham and Langford (1996) has shown that there are lagged relationships between incidence and temperature of approximately one month, suggesting that effects occurring in the food production industry may be influencing incidence of food poisoning, prior to purchase and consumption by consumers. Weekly mean temperatures were calculated from daily maximum and minimum temperatures obtained from the UK Meteorological Office for the time period studied. From modelling the weekly lags of temperature with food poisoning incidence, it was found that the effect of temperature could be divided into two effects, the first comprising the mean of this week's and the previous week's temperature ($T01$), and the second being the mean of the temperature two to five weeks in the past ($T25$) (Bentham $et\ al.$, 2000). In addition, indicator variables were included for holiday weeks such as Christmas and Easter (HOL) and the week following a holiday week ($HOL1$), since both reporting and attendance at primary health care outlets tend to follow a different pattern at these times (Bentham and Langford, 1996).

In this case, we do not have denominator populations for the response variable, and hence expected values of cases, because we have assumed that the population at risk in each county remained roughly the same over the two years included in our study. However, there was a great deal of variability in the numbers of cases occurring in each county, and it was found that a negative binomial model provided a better fit to the data than a Poisson model, which leads to a quadratic expression for the variance at level 1 (see Section 4.2). We have in this case a hierarchical model with weekly data on incidence (labelled i) nested within each county (labelled j). We fitted the following negative binomial model to the data:

$$
\begin{aligned}
O_{ij} &\sim \text{Negbin}(\pi_{ij}), \\
O_{ij} &= \pi_{ij} + e_{0ij}\pi_{ij}^{0.5} + e_{1ij}\pi_{ij} + e_{2ij}HOL_{ij}^{0.5}, \\
\log \mu_{ij} &= \beta_{0j} + \beta_2 HOL_{ij} + \beta_3 HOL1_{ij} + \beta_{4j} T01_{ij} + \beta_{5j} T25_{ij}, \\
\beta_{0j} &= \beta_0 + u_{0j}, \\
\beta_{4j} &= \beta_4 + u_{4j}, \\
\beta_{5j} &= \beta_5 + u_{5j}, \\
\begin{bmatrix} u_{0j} \\ u_{4j} \\ u_{5j} \end{bmatrix} &\sim N\left(\begin{bmatrix} 0 \\ 0 \\ 0 \end{bmatrix}, \begin{bmatrix} \sigma_{u0}^2 & \sigma_{u04} & \sigma_{u05} \\ \sigma_{u04} & \sigma_{u4}^2 & \sigma_{u45} \\ \sigma_{u05} & \sigma_{u45} & \sigma_{u5}^2 \end{bmatrix} \right), \\
\begin{bmatrix} e_{0ij} \\ e_{1ij} \\ e_{2ij} \end{bmatrix} &\sim N\left(\begin{bmatrix} 0 \\ 0 \\ 0 \end{bmatrix}, \begin{bmatrix} \sigma_{e0}^2 & \sigma_{e01} & \sigma_{e02} \\ \sigma_{e01} & \sigma_{e1}^2 & 0 \\ \sigma_{e02} & 0 & 0 \end{bmatrix} \right).
\end{aligned}
\qquad (4.8)
$$

Table 4.4 Negative binomial model of food poisoning incidence.

	PQL^2 Negbin	
	Estimate	SE
Fixed part		
β_0	1.93	0.15
$\beta_2(\times 10^{-1})$	−2.63	0.28
$\beta_3(\times 10^{-1})$	0.73	0.27
$\beta_4(\times 10^{-3})$	1.67	0.37
$\beta_5(\times 10^{-3})$	6.84	0.45
Random part		
Level 2: counties		
σ^2_{u0}	1.12	0.22
$\sigma_{u04}(\times 10^{-3})$	−1.09	0.43
$\sigma^2_{u4}(\times 10^{-6})$	1.50	1.42
$\sigma_{u05}(\times 10^{-3})$	−1.15	0.51
$\sigma_{u45}(\times 10^{-6})$	1.04	1.32
$\sigma^2_{u5}(\times 10^{-6})$	4.32	2.11
Level 1: weeks		
σ^2_{e0}	7.17	4.15
σ_{e01}	−1.03	0.99
σ^2_{e1}	0.41	0.23
σ_{e02}	−0.72	0.62

The results for this complex model are given in Table 4.4. The interpretation of the parameters is as follows, starting with the fixed part of the model. We have an overall intercept β_0, which represents an average number of cases per week in all the counties included in the study when the other fixed coefficients are zero. This is on a logarithmic scale, and is equal to $\exp(1.93) = 6.89$. The parameters β_2 and β_3 show that there are relatively fewer cases of food poisoning reported in holiday weeks, and slightly more cases reported in the weeks following holiday weeks. The parameter β_4 shows that there is a positive relationship between mean temperature for this week and the previous week with food poisoning incidence. However, although statistically significant ($p < 0.05$), this parameter is about four times smaller than β_5, the parameter that measures the effects of temperature two to five weeks previously. Hence we can deduce that, on average, temperatures two to five weeks ago are more important in predicting incidence of food poisoning in the present for a substantive discussion, (see Bentham *et al.*, 2000).

However, the relationships between incidence and temperature are not constant across all counties. This is demonstrated in the random parameter matrix estimated at level 2 of the model. In particular, there appears to be significant variation in the relationship between incidence and temperature two to five weeks ago (σ^2_{u5}). The correlation between *T25* and the intercept is also

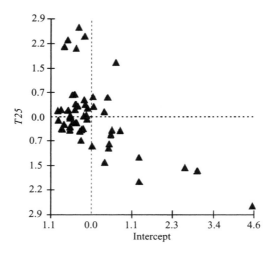

Figure 4.4 Plot of standardised residuals for the slope (*T25*) and intercept.

interesting, since it is negative (−0.53). This suggests that counties with lower numbers of cases (in general, counties with smaller populations) have more positive relationships between food poisoning incidence and temperature two to five weeks ago. This effect is shown graphically in Figure 4.4, where intercept residuals are plotted against slope residuals for *T25*. The two points at the bottom right of the diagram are the large urban centres of Greater London and West Yorkshire. Subsequent analysis of the residuals did in fact find a positive relationship between the size of effect and rurality, suggesting that there are different processes operating in urban and rural areas (Bentham *et al.*, 2000).

The level-1 Poisson part of the model also has complex variance parameters. The model is negative binomial, so parameters for both σ_{e0}^2 and σ_{e1}^2 are estimated, as well as their covariance. However, an extra parameter σ_{e02} modelling the covariance for the holiday weeks dummy variable is also included, showing that, on average, holiday weeks displayed lower variance than the other weeks included in the dataset.

4.4 DISCUSSION

We have examined two quite different datasets using a Poisson multilevel model, and found interesting features of the data and relationships that would not be evident from a simple, single-level analysis. In the first example, we showed that there was significant variation in testis cancer mortality between higher-level units such as regions and, to a degree, nations as well. This may suggest differences in reporting between different areas, but may also be linked to differences in lifestyle, culture or exposure to environmental pathogens between different regions and nations. In the second example, we found that incidence of food poisoning was related to temperature, but that these relationships were not the same for all counties over time. In particular, the relationships tended to be stronger for rural areas than for urban areas.

One of the most interesting features of the models shown is the examination of residuals at higher levels to determine if there are any unusual areas present in the data. However, our current model assumes a multivariate normal distribution of residuals at higher levels, which may not be an accurate reflection of the pattern that the residuals form. Further research is currently been undertaken into using Markov-chain Monte Carlo methods for better estimation of both fixed and random parameters. A further important feature of geographical data is the possibility of spatial autocorrelation, where areas geographically close to each other may have similar causal environmental processes operating, and hence be expected to have more similar incidence or mortality from a disease. The methods for spatial analysis are discussed in Chapter 10.

One final point is worth considering – at present, it is not possible to provide a sensible measure of overall goodness-of-fit for Poisson multilevel models. This is because we have several residuals at higher levels of the model which are not on a linear scale. However, a simulation method is provided in Langford *et al.* (1998) for estimating the deviance in a binary response model, and this could theoretically be adapted to the Poisson case, although their are technical issues to be tackled with the calculation.

CHAPTER 5

Multivariate Multilevel Models

Alice McLeod

MRC Social and Public Health Services Unit, University of Glasgow, UK

5.1 INTRODUCTION

5.1.1 Multivariate data

Multivariate data arise when more than one measurement or observation is made for each case in a sample of subjects. There are a variety of statistical methods that may be employed to analyse such data and these are classed under the broad heading of multivariate analysis. Methods such as discriminant analysis or cluster analysis are used when the purpose is to divide individuals into homogenous groups on the basis of the observed data. Principal components analysis is concerned with reducing the dimensionality of the data when the number of observations for each individual is large. Latent structure models are designed to estimate underlying (i.e. latent) variables using the observed multivariate data, and include such methods as factor analysis. Multivariate linear models are a natural extension of the univariate linear model, and provide a means of estimating associations between a set of explanatory variables and a response made up of more than one dependent variable. This means that, rather than each case having a single response, each has a vector of responses. Given its wide range of practical applications, the multivariate linear model or, more generally, multivariate regression will be the main focus of this chapter. Before going on to formulate the model within the multilevel framework, it will be useful to discuss some of the properties that multivariate regression proffers over univariate regression.

5.1.2 Multivariate regression

In studies of health, there is often more than one response variable or outcome of interest. Some measures of health are naturally multidimensional, for

Multilevel Modelling of Health Statistics Edited by A.H. Leyland and H. Goldstein

example, systolic and diastolic blood pressures and (in studies of growth) weight and height. More generally, a researcher may choose to examine more than one outcome; for example, in comparing a surgical intervention in two groups, one may be interested in both clinical outcome and patient-based outcomes such as quality of life. Indeed, in clinical trials, it is often the case that more than one outcome or endpoint must be compared between treatment groups. A common approach to such data is to carry out a univariate analysis for each response of interest and ignore the fact that measures made on the same individual are likely to be correlated. Indeed, if the response comes from a multivariate normal distribution and the data are balanced such that all individuals have been observed for each response, the coefficient estimates from a multivariate regression will be the same as those from separate univariate regressions. There are, however, a number of good reasons for using the multivariate approach. First is the issue of multiple tests and significance levels, which is particularly relevant to clinical trials. If one is interested in the effect of a particular covariate (e.g. treatment group) on p different responses then using univariate regressions results in p separate significance tests and an increased probability of finding a significant effect by chance. While it is possible to use some adjustment in order to retain an overall significance level, the multivariate approach allows each covariate effect to be tested simultaneously using a single joint test statistic (e.g. Wald's test) with a fixed type I error. The multivariate approach also provides a means of formally testing differential associations between independent variables and each outcome of interest. Assuming that the outcomes are measured on the same scale, it is possible to construct a joint test contrasting coefficient estimates whereas covariates obtained from a univariate approach may only be compared informally. Another reason for using a multivariate model is that the associations between outcomes may themselves be of interest and it will be informative to model multiple responses simultaneously in order to estimate the degree of correlation between them.

Traditional methods of multivariate regression, such as ordinary least squares (OLS), have been limited by a number of factors that may have led to their infrequent use; Pocock (1983) points out that when dealing with multiple endpoints in clinical trials 'methods of multivariate analysis... are often difficult to apply or interpret'. Such methods are subject to the restriction of having a balanced design and assumptions regarding the homogeneity of the covariance structure for the observed data. If only those cases with complete response vectors may be included and there is differential missingness across responses then a multivariate analysis may lead to a serious loss of cases and power compared with separate univariate regressions. The issue of a homogenous error structure is also made more complex in the multivariate case by the fact that variation in the sample is described not by one term but, with a p-dimensional response, by $p(p + 1)/2$ variance and covariance terms. Perhaps the most important restriction of the OLS approach has been that applications are limited to multivariate normal data, and such data are not always common in the field of health. More often, researchers will be faced by discrete responses or combinations of discrete and continuous responses. This chapter discusses how each of these limitations may be overcome with the use of multilevel models, and illustrates the methods using a variety of health data.

5.2   THE MULTIVARIATE LINEAR MODEL

5.2.1   Basic notation

In the context of multilevel models, multivariate data can be considered in a similar way to repeated measures, because observations are nested within individuals. However, rather than the same variable recorded at different time points, a multivariate response design has different variables recorded on one occasion. The following formulation of the multivariate linear model is taken from Goldstein (1995) with a slight amendment to notation. We assume that we have n realisations of a multivariate random vector $\mathbf{y}$ of dimension p, i.e. p measurements or observations made on each of n individuals, and that

$$\mathbf{y} \sim N(\boldsymbol{\beta}, \boldsymbol{\Sigma}),$$

where $\boldsymbol{\beta}$ is a mean vector of dimension p and $\boldsymbol{\Sigma}$ is a $p \times p$ symmetric covariance matrix, such that the diagonal elements are the variances of the p responses and the off-diagonal elements the covariances between responses. A simple model may be written as

$$y_{ij} = \sum_t \beta_{0tj} z_{ijt}, \qquad t = 1, \ldots, p, \tag{5.1}$$

where

$$z_{ijt} = \begin{cases} 1 & \text{for } t = i, \\ 0 & \text{otherwise.} \end{cases}$$

Thus y_{ij} is the ith response for the jth individual, and t indexes a set of p measurements. The zs are dummy variables used to distinguish between each of the p responses. Further, the mean may be written as

$$\beta_{0ij} = \beta_{0i} + e_{0ij}, \tag{5.2}$$

such that

$$\text{var}(e_{0ij}) = \sigma^2_{e,\,i}$$

and

$$\text{cov}(e_{0ij}, e_{0kj}) = \sigma_{e,\,ik}.$$

Writing the model in such a way, it is easy to see the multivariate linear model as a two-level multilevel model with measurements at level 1 nested within individuals at level 2. No parameters are estimated at level 1, and this lowest level is used simply to define the multivariate structure. In equation (5.2), the parameter β_{0i} provides an estimator for the ith component of the mean vector

β, and the random effects e_{0ij} are the residuals of the observed data around the mean of the ith response. The variation in the level-2 random effects, i.e. the residuals, then provide estimates of the components of Σ, namely, the between-subject variances $\sigma_{e,i}^2$ and covariances $\sigma_{e,ik}$.

More generally, it will be of interest to extend equation (5.1) to incorporate a set of q explanatory variables, x_{qj}, in the linear predictor of y such that

$$y_{ij} = \sum_t \beta_{0tj} z_{ijt} + \sum_t \sum_q \beta_{qt} x_{qj} z_{ijt}. \qquad (5.3)$$

This parameterisation allows each response to have its own set of coefficients for the explanatory variables x_{qj}, and while equation (5.3) implies that each response has the same set of explanatories, this need not be the case in practice.

Note that in equations (5.1) and (5.3), the covariance matrix is assumed to be constant across the whole sample. This is equivalent to the common assumption of homoskedasticity in the univariate linear model. However, just as the multilevel framework allows explicit modelling of departures from constant variance at level 1, the multilevel multivariate model allows us to explore non-constant covariance matrices. The same principles that apply to modelling complex variation in a univariate response also apply in the multivariate case, although there is the added complexity of non-constant covariance to be considered. The following section uses a simple example to illustrate the multi-variate linear model, and includes an exploration of non-constant covariance.

5.2.2 Example 1: blood pressure in a low-birthweight cohort

The following data come from the most recent phase of a Scottish cohort study of very-low-birthweight (VLBW, $<1500\,\text{g}$) children and their classroom peers designed to investigate various aspects of health and development in these children at eight to nine years of age (McLeod *et al.*, 1996). In recent years there has been considerable interest in the fetal origins of adult health, and one focal point has been the inverse relationship of birthweight to blood pressure (BP) in later life. Current weight is known to be positively associated with BP; however, children weighing below average at birth have above average BP in adulthood. The mechanisms behind this association are unknown, and will not be explored here. Our aim is simply to establish whether the VLBW cohort is experiencing an increase in their BP at age eight to nine years relative to their classroom peers.

BP is measured by two readings, systolic (SBP) and diastolic (DBP), both measured on the same scale (mmHg), with SBP being the higher of the two. The two measures of BP are known to be highly correlated, and individuals with a high SBP are also likely to have increased DBP. Despite this association, BP is often confined to two separate univariate analyses. In the following example, we shall consider the two responses simultaneously in a multivariate linear model, and this will enable a single significance test of the main hypothesis using the joint parameter estimates and their estimated covariance matrix.

Table 5.1 Parameter estimates from two multivariate linear models for systolic blood pressure (SBP) and diastolic blood pressure (DBP) in very-low-birthweight (VLBW) children and their classroom peers.

	Model A estimate (SE)	Model B estimate (SE)
SBP		
Constant	109.7 (0.45)	91.4 (1.95)
VLBW	1.9 (0.78)	3.9 (0.77)
Weight	—	0.63 (0.06)
Females	—	−0.40 (0.70)
DBP		
Constant	57.3 (0.40)	50.8 (1.78)
VLBW	1.1 (0.68)	1.7 (0.70)
Weight	—	0.19 (0.06)
Females	—	1.8 (0.64)
σ^2_{SBP}	120.8	109.2
σ^2_{DBP}	92.4	90.4
$\sigma_{SBP, DBP}$	47.6	44.0

Table 5.1 contains results from two simple multivariate models. In the first, model A, only an intercept and a coefficient for the VLBW group are fitted for each response. The intercept provides the mean for the control group, while the VLBW coefficient provides the expected change in mean BP for children of low birthweight. While both VLBW coefficients are positive, indicating increased BP in the VLBW cohort, a joint significance test does not reject the hypothesis of equal mean vectors for the two study groups ($\chi^2 = 5.96$ on 2 df). Model B adjusts for two variables, sex and weight, both of which are known to be related to BP. The sampling design has ensured that the two study groups are balanced by sex; however, the VLBW cohort differ considerably from their classroom peers in terms of weight and are over 3 kg lighter on average. As expected, weight is positively associated with both blood pressures; the association is stronger for SBP than for DBP. SBP would not appear to be different for boys and girls, although girls seem to have a higher DBP than boys. Most importantly, by adjusting for weight, the difference in blood pressure between the two study groups has increased from model A; on average, the VLBW cohort experience an increase of 3.9 mmHg in their SBP and 1.7 mmHg in their DBP. This shows that the VLBW effects in model A were confounded by different current weights in the two samples. The joint test for the VLBW effect in model B is significant ($\chi^2 = 25.24$ on 2 df), and we can conclude that the VLBW children do have increased blood pressure compared to their classroom peers.

Also provided in Table 5.1 are estimates of the SBP and DBP variances and the covariance between the two. The variation in each response decreases slightly from model A to model B as more explanatory variables are added.

Table 5.2   Non-constant covariance for blood pressure in males and females.

	Males	Females
σ^2_{SBP}	107.6	110.9
σ^2_{DBP}	89.7	91.0
$\sigma_{SBP,DBP}$	36.7	51.2

The correlation between the two blood pressures can be calculated directly from the variance and covariance estimates using the formula

$$\rho_{SBP,DBP} = \frac{\sigma_{SBP,DBP}}{\sqrt{\sigma^2_{SBP}\sigma^2_{DBP}}}. \tag{5.4}$$

Not surprisingly, SBP and DBP are positively correlated (0.45), although the correlation is smaller than that observed in adults owing to the age at which the children have been measured.

Models A and B assume a constant variance–covariance matrix for all children in the study; however, we can investigate if this is different for different groups. Standard multivariate regression (e.g. MANOVA) assumes constant variance–covariance for the entire sample – in the same way that ANOVA assumes constant variance. In an analogous way to exploring non-constant variance at level 1, the multilevel framework allows exploration of non-constant covariance matrices for a multivariate response. To illustrate this, we shall consider different covariance matrices for girls and boys. In the multilevel approach this means a reparameterisation of the variance at level 2 using dummy indicators for sex; this may be done using the constants and the female dummies in an analogous way to modelling complex level-1 variation for a univariate response (see Chapter 1, Section 1.9) or by creating four new dummy variables, one for each response for both boys and girls. Note that it is also possible to use continuous explanatory variables to explore the level-2 covariance structure. Table 5.2 contains estimates for separate covariance matrices by sex obtained from a third model. The between-girl variations in SBP and DBP do not appear to be very different from those of the boys. However, the covariance, and consequently the correlation between SBP and DBP, is greater for the girls: 0.51 compared with 0.37 for boys. This finding may be a result of girls reaching a pre-pubertal growth spurt prior to boys, which in turn causes a sequential increase in SBP followed by DBP, and results in the two measures of blood pressure being more highly correlated.

5.3   EXTENSIONS OF THE MULTIVARIATE LINEAR MODEL

5.3.1   Clustered multivariate responses

It will sometimes be the case that the individuals generating the multivariate responses are themselves clustered in some way and that this clustering should be incorporated in the model. For example, individuals on which a multivariate response is measured may be clustered within organisational or geographical groups such as hospitals, general practices or areas. Within the context of clinical trials, a multilevel multivariate model may be appropriate for either a meta-analysis of trials (individuals within trials) or a multicentre trial (individuals within centres) in which multiple outcomes are recorded. A slightly different example arises in cohort studies, where generally a number of measurements are made on each individual at each follow-up assessment and, depending on the research question, it may be appropriate to analyse the data as multivariate repeated measures. This last example differs from those above in which responses (level 1) are clustered within individuals (level 2), who are themselves clustered within higher-level units (level 3+). The multivariate repeated measures design leads to responses (level 1), clustered within follow-up phase (level 2), clustered within individuals (level 3). Incorporating more complex multivariate regression designs is straightforward using the multilevel approach; the majority of analytical designs leading to a multilevel model may be extended to incorporate a multivariate response simply by adding an extra level at the bottom of the hierarchy. Equation (5.3) can be extended to a three-level model such that

$$y_{ijk} = \sum_t \beta_{0tjk} z_{ijkt} + \sum_t \sum_q \beta_{qt} x_{qjk} z_{ijkt}, \tag{5.5}$$

where y is the ith response for the jth individual in the kth cluster and the zs are dummies as before. Here

$$\beta_{0ijk} = \beta_{0i} + e_{0ijk} + u_{0ik}, \tag{5.6}$$

and the between-cluster variance for the ith response is given by

$$\text{var}(u_{0ik}) = \sigma_{u,\,i}^2,$$

while the between-cluster covariance for the ith and rth responses is given by

$$\text{cov}(u_{0ik}, u_{0rk}) = \sigma_{u,\,ir}.$$

These estimates can then be used to calculate the degree of correlation between responses among higher-level units.

5.3.2 Discrete outcomes

Up to this point, the discussion has focused on linear models appropriate for continuous, normally distributed responses; however, the multilevel formulation and various estimation algorithms (e.g. IGLS) also allow the analysis of multivariate discrete responses. Thus, for multiple binary outcomes, or multiple proportions, one can carry out multivariate logistic regression, and for multiple counts, multivariate Poisson regression. To illustrate this, consider a response vector $\mathbf{y}$ containing p binary responses. We can write

$$y_{ij} = \pi_{ij} + e_{ij}, \tag{5.7}$$

where π_{ij} is the probability that the ith response for the jth individual is equal to 1 and e_{ij} is the residual at the individual level such that

$$E(e_{ij}) = 0, \text{var}(e_{ij}) = \pi_{ij}(1 - \pi_{ij})\sigma_i^2.$$

As with a univariate logistic regression, σ_i^2 is the scale parameter, which is included in the model to investigate over- or under-dispersion in the ith response. This parameter is estimated, and if it is equal to one then the assumption of binomial variation is reasonable.

To estimate the coefficients of q dependent variables associated with the probability of each of the p outcomes, a generalised linear model with a logit link may be used, and is written as

$$\text{logit}(\pi_{ij}) = \log\left(\frac{\pi_{ij}}{1 - \pi_{ij}}\right) = \sum_t \beta_{0tj} z_{ijt} + \sum_t \sum_q \beta_{qt} x_{qj} z_{ijt}. \tag{5.8}$$

Note that the linear predictor in (5.8) is identical to that in (5.3), implying that models for multivariate discrete responses are constructed in the same way as those for multivariate normal responses. As with univariate analyses, it is only in terms of estimation and interpretation of the resulting coefficients that the two situations differ. Combining equations (5.7) and (5.8),

$$y_{ij} = \pi_{ij} + e_{ij} = \left\{1 + \exp\left[-\left(\sum_t \beta_{0t} z_{ijt} + \sum_t \sum_q \beta_{qt} x_{qj} z_{ijt}\right)\right]\right\}^{-1} + e_{ij}. \tag{5.9}$$

The coefficient for the qth explanatory may be interpreted as the increase in log odds associated with a unit increase in x_{qj}. Slightly more care must be taken with interpretation of the parameters in the random part of the model. If the assumption of binomial variation is true, the estimate of covariance at level 2 has the form

$$\text{cov}(e_{1j}, e_{2j}) = \hat{\pi}_{(11)j} - \hat{\pi}_{1j}\hat{\pi}_{2j}, \tag{5.10}$$

where $\hat{\pi}_{(11)j}$ is the predicted probability of a positive response for both outcomes (Goldstein, 1995). While this covariance term does not offer a necessarily meaningful interpretation itself, it can be used with the predicted probabilities from the fixed part of the model, $\hat{\pi}_{1j}$ and $\hat{\pi}_{2j}$, to derive an estimate of the more easily interpreted $\hat{\pi}_{(11)j}$.

As with a multivariate normal response, the model described above may be extended to more complex models incorporating clustering at higher levels. Equation (5.8) may be extended to include a third level such that

$$\text{logit}(\pi_{ijk}) = \sum_t \beta_{0tjk} z_{ijkt} + \sum_t \sum_q \beta_{qt} x_{qjk} z_{ijkt}, \tag{5.11}$$

where

$$\beta_{0ijk} = \beta_{0i} + u_{0ik}.$$

As with a univariate discrete response, higher-level residuals are assumed to be normally distributed, making interpretation of higher-level variances and covariances straightforward.

5.3.3 Mixed discrete and continuous outcomes

Multivariate responses need not be measured on the same scale or even take the same numerical form. For example, in measuring outcomes from surgery, one may have measured the patient's response to the intervention using a number of assessments, some of which produce an interpretable continuous score and some of which may be classed into a satisfactory/unsatisfactory outcome. To illustrate how such models are constructed and how they may be interpreted, let us assume a bivariate response with one normally distributed outcome and one binary outcome. Further assume that these multivariate observations have been made on a group of individuals who are themselves clustered within higher-level units. A model for such a response may be written as

$$y_{ijk} = z_{ijk} \left\{ 1 + \exp\left[-\left(\beta_{01} + \sum_q \beta_{q1} x_{q1jk} + u_{1k} \right) \right] \right\}^{-1}$$

$$+ z_{ijk} e_{1jk} + (1 - z_{ijk}) \left(\beta_{02} + \sum_q \beta_{q2} x_{q2jk} + u_{2k} + e_{2jk} \right),$$

(5.12)

where $z_{ijk} = 1$ if the response is binary and $z_{ijk} = 0$ if the response is continuous. The es are the individual level residuals and it is assumed that

$$E(e_{1jk}) = 0, \quad \mathrm{var}(e_{1jk}) = \pi_{1jk}(1 - \pi_{1jk})\sigma_1^2,$$

while

$$E(e_{2jk}) = 0, \quad \mathrm{var}(e_{2jk}) = \sigma_2^2.$$

The us are the higher-level residuals (level 3), and are assumed to be normally distributed with zero mean and a joint covariance matrix whose components are estimated from the data. The interpretation of the above parameters is the same as those from standard linear and logistic regression models, and it is only the covariance term at level 2 that must be considered more carefully. The individual-level covariance term is given by

$$\mathrm{cov}(e_{1jk}, e_{2jk}) = (1 - \hat{\pi}_{1jk})\hat{\pi}_{1jk}\left(y_{2jk}^1 - y_{2jk}^0 \right),$$

(5.13)

where $\hat{\pi}_{1jk}$ is the predicted probability that $y_{1jk} = 1$, and y_{2jk}^1 and y_{2jk}^0 are the predicted values of the continuous response for $y_{1jk} = 1$ and $y_{1jk} = 0$ respectively (Goldstein, 1995). This means that a positive covariance may be interpreted as an increase in the continuous response for individuals with positive values of the binary response, and a negative covariance as a decrease in the continuous response for individuals with positive values of the binary response. That is, the covariance from this model may be interpreted in a similar way to that between two continuous responses. It is worth noting that it is relatively straightforward to estimate mixed multivariate response models in MLwiN when dealing with either a mix of continuous and dichotomous/proportion outcomes or a mix of continuous and count outcomes. However, there are no currently available facilities for estimating a mixture of discrete variables (e.g. binary and counts), although, in theory, such models may be constructed. The following example illustrates this approach and applies the model in equation (5.12) to a sample of routine hospital data.

5.3.4   Example 2: LOS and readmission rates

This example involves two measures of hospital utilisation: length of hospital stay in days (LOS) and emergency readmissions within 28 days of discharge. Both the average length of stay and readmission rates are known to vary considerably between hospitals, even after adjustment for case-mix. Variations in LOS are important because this measure correlates strongly with the cost of a hospital episode, while variations in emergency readmission rates are important because this is used as an indicator of the quality of care received prior to readmission. LOS has been decreasing steadily over the years, and it has been hypothesised that shorter stays may be associated with an increased risk of readmission. Previous studies have investigated this hypothesis by comparing adjusted average LOS and rates of readmission between hospitals. By using a multilevel multivariate model, however, it is possible to estimate simultaneously the association between these two measures at both patient and hospital level.

This example is taken from a larger project designed to investigate LOS and readmission rates for a number of diagnostic and surgical patient groups using the King's Fund Comparative Database (McCulloch *et al.*, 1997). Here we consider a sample of 40 471 patients discharged with chronic obstructive airways disease (COAD) from 98 hospitals during a period of 17 months. A total of 3078 (7.6%) patients were readmitted as an emergency within four weeks of discharge, and the average LOS was just over ten days. While LOS can be considered a continuous variable, it is often the case that its distribution is skewed to the left and, in this example, the logarithm of LOS was taken to reduce the degree of skewness.

Table 5.3 details a selection of fixed effects found to be significantly associated with LOS and emergency readmission; covariate effects are expressed as the factor change in LOS and the odds ratio (OR) of readmission relative to a baseline category shown in parentheses. Thus, relative to elective admissions, patients admitted as an emergency stayed twice as long in hospital and were

Table 5.3   Parameter estimates from a multilevel multivariate model of LOS and emergency readmission.

Covariate (baseline)		LOS (factor increase)	Emergency readmission (OR)
Age (70 years)	*10-year increase*	1.18	1.10
Sex (male)	*female*	1.09	0.84
Admission (elective)	*emergency*	2.03	2.07
Type of surgery (none)	*open operations on bronchus*	0.36	0.32
	other operations	1.33	—
Discharge (alive)	*dead at discharge*	0.70	—

Table 5.4 Variance estimates for a multilevel multivariate model of length of stay (LOS) and emergency readmission.

Random parameter	Estimate (SE)	Correlation
Hospital level		
$\sigma^2_{u,\,READM}$	0.211 (0.037)	
$\sigma_{u,\,READM,\,LOS}$	0.028 (0.009)	0.404
$\sigma^2_{u,\,LOS}$	0.023 (0.004)	
Patient level		
$\sigma^2_{e,\,READM}$	1 (—)	
$\sigma_{e,\,READM,\,LOS}$	0.022 (0.004)	—
$\sigma^2_{e,\,LOS}$	0.746 (0.005)	

twice as likely to be readmitted. Also, patients undergoing an operation on the bronchus had shorter stays than patients for whom no operation was performed, and were also less likely to be readmitted. Note that OR estimates are missing for patients who had an alternative operation and those who were dead at discharge. The estimate is missing for the first group because there were insufficient numbers readmitted to estimate this coefficient, while for those who were dead at discharge, the response is missing by default, i.e. these patients could not be readmitted at a later date because they died during the index episode. This illustrates that not all responses need to have the same set of explanatory variables included in a model.

The random coefficients from this model are detailed in Table 5.4. Variation in LOS is much greater between patients than between hospitals, although the latter is still significant. There is also a great deal of unexplained variation between hospitals in terms of readmission rates. The covariance between each response is positive at both patient and hospital levels. By plugging in the estimated covariance and the average readmission rate (7.6%) into equation (5.13), and recalling that log(LOS) was used in the fitted model, we find that patients who were readmitted stayed, on average, 1.4 days longer than those who were not readmitted. The application of equation (5.4) to calculate the correlation coefficient is inappropriate at the individual level because of the mixture of distributions. However, because the level-3 residuals are assumed to come from a multivariate normal distribution, the correlation at the hospital level may be interpreted directly, and shows that hospitals with longer than average LOS also have above average readmission rates – a relationship that contradicts the original hypothesis.

5.4 MISSING DATA

5.4.1 Responses missing at random

The multilevel approach to repeated measures analysis is often preferred because of its ability to handle missing responses efficiently. An analogous argument may be used for the multilevel approach to multivariate regression, because the method allows the inclusion of individuals who have not been measured or who have not responded to all outcomes of interest. The application of the method to incomplete response vectors is straightforward when it is assumed that responses are missing at random. This principle assumes that individuals with incomplete response vectors may be included in the analysis on the basis that the association between their responses will on average mimic that which is observed for individuals with complete response vectors. This feature of the multilevel approach is very attractive when compared with the standard OLS approach, which omits all cases with incomplete response vectors and may sometimes lead to a serious reduction in sample size and power.

5.4.2 Responses missing by design

One way to minimise missing data in questionnaires is to reduce the number of questions that the respondent is expected to answer. Assuming that p responses are of interest and that the data will be analysed using multivariate regression, it is possible to design the study such that each individual is only required to provide responses to a subset of the total p responses. Assuming that each combination of responses is replicated sufficiently, it is then possible to estimate the association between all responses using this reduced data design. Such designs are called rotation designs, and are effectively creating datasets that have data missing at random and then using a method that can handle such unbalanced data to compute the estimates of interest. Further discussion of this approach and an application can be found in Goldstein (1995).

5.4.3 Structural missing responses

The ability of the multilevel approach to handle missing data leads to an interesting application of the method. In many situations, incomplete response vectors exist, but the incompleteness cannot be considered random – rather it is an artefact of the way in which the data or the research question is structured. A simple example of this is provided by behavioural outcomes such as smoking or drinking, which may be measured qualitatively (smokes/does not smoke) and quantitatively (smokes y cigarettes per day). Traditionally, this outcome may be analysed on a single scale, with non-smokers being assigned a zero for the number of cigarettes smoked per day. Not only is this approach unsatisfactory from a theoretical point of view, it can also create the problem of bimodality in the resulting distribution, with a peak at zero and a peak at the average number of cigarettes smoked. An alternative way to analyse such an outcome is to use a mixed discrete–continuous response model as described in Section 5.3.3, in which the dichotomous response y_{1jk} indicates smoking status (non-smoker=0, smoker=1) and the continuous response y_{2jk} measures the number of cigarettes smoked per day. Structuring the model in this way means that y_{2jk} is defined as 'missing' for all $y_{1jk} = 0$. The data can no longer be considered multivariate at the individual level, because one response is conditional on the other; however, if the data are nested, within areas for example, the correlation between responses may be estimated at levels higher than the individual. Duncan (1997) applies this method to investigate smoking behaviour and, specifically, whether areas with higher rates of smoking impact upon the amount smoked by individuals resident in those areas. This relationship is quantified by considering the direction and magnitude of the area–level covariance between smoking behaviour and cigarette consumption. The following example illustrates a model with structural missingness for two dichotomous responses, in which each outcome measures death in two different settings.

5.4.4 Example 3: Hospital and non-hospital deaths from AMI

Death in hospital or in the 30 days following discharge is used as a measure of hospital performance. One of the problems with such an outcome is that it fails to take account of those deaths that occur away from the individual reaching hospital; for some conditions, such as acute myocardial infarction (AMI), the number of non-hospital deaths may be substantial. Deaths before reaching hospital are likely to be the more severe cases and the probability of these cases reaching hospital will vary from area to area. For example, a person living close to a hospital in an urban area may be more likely to get to the hospital than a person living further away in a rural area. Thus, the characteristics of an area and the persons living there will be reflected in the 'performance' of the hospitals serving those areas. This example attempts to take account of non-hospital deaths by defining two variables: (i) death from AMI before reaching hospital and (ii) death from any cause, in hospital or in the 30 days following discharge, after being admitted with AMI. This example is a simplified version of that reported by Leyland and Boddy (1998), and concentrates solely on individuals and areas, omitting the level of hospital for the purpose of illustration. The sample is defined as all individuals resident in Scotland who experienced an identified AMI in 1993, either through death or hospital treatment. The data were derived from Scotland's linked routine hospital discharge records and death records.

The problem has been formulated into a bivariate dichotomous response for individuals nested within areas; Table 5.5 shows each combination of outcomes together with the number of individuals in each category. Clearly, $y_{1jk} = 0$ for all individuals who reached hospital, except for a small group ($n = 557$) who experienced an AMI while already being treated in hospital – this group could not die from an AMI before reaching hospital, so y_{1jk} is defined as missing. For those individuals who died before reaching hospital, y_{2jk} is defined as missing.

Table 5.6 contains estimates of a selection of variables that were used to explore variations in each outcome. Note that deprivation is measured at the area level using Carstairs scores based on the 1991 census, and areas have been grouped into quartiles such that the lowest quartile contains the most affluent areas and the upper quartile the most deprived. As a result of both responses being dichotomous, the covariate coefficients, expressed as odds ratios (OR), are directly comparable between each response. Death before reaching hospital

Table 5.5 Outcomes for hospital and non-hospital deaths for acute myocardial infarction (AMI).

Outcome	y_{1jk}: death from AMI before reaching hospital	y_{2jk}: death after hospital treatment for AMI	n
Admitted to hospital with AMI and did not die during admission or within 30 days of discharge	0	0	10 771
Admitted to hospital with AMI and died during admission or within 30 days of discharge	0	1	3 031
Died from AMI without reaching hospital	1	—	9 391
Experienced AMI during admission to hospital for other complaint	—	0 or 1	557

Table 5.6 Parameter estimates for hospital and non-hospital deaths for AMI (estimates in *italics* significant at 0.05 level).

Covariate (baseline)		Death from AMI before reaching hospital (OR)	Death after hospital treatment for AMI (OR)
Age (70 years)	*10-year increase*	*1.50*	*2.13*
Sex (male)	*female*	*0.71*	1.02
Previous AMI (no)	*yes*	*0.80*	0.92
Deprivation (most affluent)	*most deprived*	1.04	*1.21*

was significantly associated with age, sex and whether or not the individual had a previous AMI. A slightly different set of results was found for death after being admitted for treatment of AMI; age was significantly associated with death but sex was not, the effect of a previous AMI was less pronounced and non-significant, and the probability of death was significantly increased in more deprived areas.

Table 5.7 Area-level variation in non-hospital and hospital deaths for AMI.

Random parameter	Estimate (SE)	Correlation
Area level		
$\sigma^2_{u, NONHOSP}$	0.085 (0.012)	
$\sigma_{u, NONHOSP, HOSP}$	0.038 (0.010)	1.16
$\sigma^2_{u, HOSP}$	0.013 (0.015)	

Table 5.7 details the variation between areas in terms of death before and after reaching hospital – there is considerably more variation in non-hospital deaths than there is in deaths following admission; indeed, the between area variation in deaths following admission would not appear to be significant. The covariance term is positive and significant; however, we must exercise some caution when interpreting this parameter given that the variation in y_{2jk} is close to zero. As Duncan (1997) points out, the covariance must be considered together with the variances in order to understand any association between the two outcomes. The fact that the variation around the intercept for deaths following admission is close to zero suggests that most of the variation between areas has been explained by the variables included in the fixed part of the model. Therefore it does not necessarily make sense to interpret the positive covariance as an indication of correlated area effects when there is a negligible effect of area on one of the outcomes. It is interesting to note that the correlation calculated using equation (5.4) is outside the range of possible values. This reflects the fact that the variance and covariance estimates are simply point estimates of parameters calculated from a regression on the residuals; therefore the correlation coefficient derived from these point estimates leads to a value greater than one.

CHAPTER 6

Outliers, Robustness and the Detection of Discrepant Data

Toby Lewis
School of Mathematics, University of East Anglia, Norwich, UK

Ian H. Langford
CSERGE, School of Environmental Sciences, University of East Anglia, Norwich, UK

6.1 INTRODUCTION

We begin with a general description of the problems considered in this chapter. Methods of dealing with them are then illustrated by carrying through a detailed analysis of a data set on cancer mortality.

In the typical multilevel analysis situation we are discussing, we have data on some phenomenon of interest (e.g. mortality rates or duration of illness), and we want to investigate possible connections of this phenomenon with various factors (e.g. age of patient or local atmospheric pollution). Our measures of these factors are the explanatory variables, and our measure of the phenomenon is the response. We look for evidence of any relationship between the response and the explanatory variables that could reasonably account for or 'explain' the observed variation in the response values. For this purpose, we set up a statistical model and investigate whether or not it adequately fits the data set. If it does, we have plausible evidence of relationship; if not, we may modify or may change the model, or we may conclude that no relationship exists.

Suppose we are analysing a data set in terms of a chosen model. We may find that the fitted parameters of the model depend to an exceptional extent on one or more individual data points or units. In this chapter, we discuss how to identify such points, how to take account of them in analysing the data, and how to use them to throw light on the underlying structure of the data and the phenomenon under investigation.

Multilevel Modelling of Health Statistics Edited by A.H. Leyland and H. Goldstein
©2001 John Wiley & Sons, Ltd

The most obvious kind of 'special' data point is an outlier or outlying point. This is a data point which appears to be out of line with the rest of the data, in other words inconsistent with the model fitting the rest of the data. We illustrate this in Figure 6.1 in the simple context of a linear regression. The points P within the dotted boundary are reasonably fitted by a linear regression model. If we had the further data point A, it would be an outlier in the data set $P + A$, relative to the linear regression model. It has a large residual about the fitted line; equivalently, its inclusion in the data set significantly increases the residual variation.

If instead of A we add the data point B to give the data set $P + B$, B is not an outlier (an outlying data point) relative to the linear regression model. But it does have a substantial effect on the fit of the model, inasmuch as its inclusion substantially increases the precision with which the slope of the regression line is estimated. This is because its x-value is at some distance from the x-values of the points P; in fact, its x-value is an outlier in the set of explanatory values for $P + B$. We say that B is a point of high leverage. Note that leverage is entirely a feature of the explanatory variable or variables, and not of the response. Thus if instead of B we had the data point C (with the same x-value as B) to give the data set $P + C$, C has the same leverage as B had. However, in contrast with B, the inclusion of C in the data set substantially changes the estimate of the regression line. We say that C is an influential point in the data set $P + C$. C is also an outlier (relative, as always here, to the linear regression model). In the data set $P + A$, the outlying point A has low leverage, but is an influential point for the intercept while not for the slope.

For simplicity we have introduced the concepts of outliers, leverage, and influential data in terms of individual data points. But, as we shall see, these concepts apply equally well to groups or subsets of points. For example, in Figure 6.1, the outlying pair of points (A, D) are influential for the slope while not for the intercept.

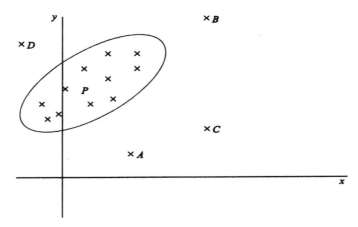

Figure 6.1   Illustration of outliers, leverage and influence in a regression analysis.

We now show how to deal with these features in the context of a particular data set from the health field, and how to use them to get a better understanding of the phenomena under review.

6.2 CANCER MORTALITY DATA

Our data set is concerned with mortality from prostate cancer in nine European countries, how it may be related to the incidence of ultraviolet light in the B band and other possible explanatory factors, and how such a relationship may differ from country to country, from region to region within a country, and from area to area within a region.

The data come from the *Atlas of Cancer Mortality in the European Economic Community* (Smans *et al.*, 1992). They cover the period 1975–80 in nine European countries. In the data set, each country is divided into regions, and each region into areas or localities; the nine countries comprise altogether 79 regions and 353 areas (see Table 6.1). To illustrate what is implied by a 'region' and an 'area', we give in Table 6.2 the breakdown into regions and areas for the UK. Clearly in the UK context, an area is essentially a county or shire. In France, it is essentially a département, and so on.

The *Atlas* gives extensive data on age-standardised cancer mortality rates for cancers relating to 14 different sites or causes (oesophagus, lung, leukaemias, urinary organs, etc). In this chapter, we use the data on just one type of cancer, namely cancer of the prostate. The number of individual prostate cancer cases covered by the data set is 284 331. The age-standardized mortality rates from prostate cancer are given for each of the 353 areas; that is our response variable, which we denote y_{ijk} for area ijk in region jk in country k. We thus have a three-level nested data set with areas, regions and countries as level-1, level-2 and level-3 units respectively. We also have the values of the following explanatory variables: x_{1ijk} is the ultraviolet index (incidence of ultraviolet light in the B band) for area ijk in region jk in country

Table 6.1 Area distributions by country.

Country	Number of regions	Number of areas
Belgium	3	11
Germany (FR)	11	30
Denmark	3	14
France	22	94
UK	11	69
Italy	20	95
Ireland	4	26
Luxembourg	1	3
Netherlands	4	11
TOTAL	79	353

Table 6.2 Regions and areas in the UK.

East Anglia	*East Midlands*	*Northern*	*North West*
Cambridgeshire	Derbyshire	Tyne & Wear	Greater Manchester
Norfolk	Leicestershire	Cleveland	Merseyside
Suffolk	Lincolnshire	Cumbria	Cheshire
	Northamptonshire	Durham	Lancashire
	Nottinghamshire	Northumberland	
South East	*South West*	*West Midlands*	*Yorkshire and*
Greater London	Avon	West Midlands	*Humberside*
Bedfordshire	Cornwall	Hereford & Worcester	South Yorkshire
Berkshire	Devon	Salop	West Yorkshire
Buckinghamshire	Dorset	Staffordshire	Humberside
East Sussex	Gloucestershire	Warwickshire	North Yorkshire
Essex	Somerset		
Hampshire	Wiltshire		
Hertfordshire			
Isle of Wight			
Kent			
Oxfordshire			
Surrey			
West Sussex			
Wales	*Scotland*	*Northern Ireland*	
Clwyd	Highland	Eastern	
Dyfed	Grampian	Northern	
Gwent	Tayside	Southern	
Gwynedd	Fife	Western	
Mid-Glamorgan	Lothian		
Powys	Borders		
South Glamorgan	Central		
West Glamorgan	Strathclyde		
	Dumfries &		
	Galloway		
	Shetland		
	Western Isles		

k, x_{2jk} is the income per capita for region jk in country k, x_{3jk} is the population density for region jk in country k, $x_0 = 1$ is a constant whose coefficient gives the intercept.

The first step is to set up a plausible model; we refer to this as an initial model, in the sense that we may need to revise it in the light of the results of the analysis. For medical reasons, there is reason to expect an association between prostate cancer and incidence of ultraviolet light, which may vary from country to country; we therefore modelled the ultraviolet index x_{1ijk} to have a random coefficient at level 3 (country). Investigation of the data indicated that variation in the relationships of prostate cancer mortality to income per capita (x_{2jk}) and population density (x_{3jk}), to the extent that it existed, was not strong; we therefore modelled the coefficients of these two variables as fixed.

6.3   AN INITIAL MODEL

We have the following initial model, which we call M1:

$$y_{ijk} = \beta_{0ijk} x_0 + \beta_{1k} x_{1ijk} + \beta_2 x_{2jk} + \beta_3 x_{3jk}, \tag{6.1}$$

$$\beta_{0ijk} = \beta_0 + v_{0k} + u_{0jk} + e_{0ijk}, \tag{6.2}$$

$$\beta_{1k} = \beta_1 + v_{1k}. \tag{6.3}$$

For the moment we shall assume that all the random variables in (6.2) and (6.3) have normal distributions. Later, we shall see how this assumption may be checked.

The estimates of the coefficients are as follows (each estimate is accompanied by its estimated standard error in parantheses):

$$\left.\begin{array}{ll} \hat{\beta}_0 = 14.71\,(0.33), & \hat{\beta}_1 = -0.160\,(0.136), \\[2mm] \hat{\beta}_2 = 0.0033\,(0.0073), & \hat{\beta}_3 = 0.00015\,(0.00027). \end{array}\right\} \tag{6.4}$$

The estimates of the random parameters (likewise with their estimated standard errors) are as follows:

$$\left.\begin{array}{l} \text{level-3 var}\begin{pmatrix} v_{0k} \\ v_{1k} \end{pmatrix} = \begin{pmatrix} 0.181\,(0.270) & \\ 0.028\,(0.111) & 0.113\,(0.071) \end{pmatrix}, \\[4mm] \text{level-2 var}(u_{0jk}) = 0.346\,(0.173), \\[2mm] \text{level-1 var}(e_{0ijk}) = 2.663\,(0.224). \end{array}\right\} \tag{6.5}$$

The deviance (-2 log likelihood), which we shall call $L1$, is equal to 1407.01.

The variance (0.113) of β_{1k}, the coefficient of the variable ultraviolet index, is less than twice its estimated standard error (0.071), which might suggest that there is no point in modelling this variable with a random coefficient at level 3, and that we should simply model its coefficient as a fixed parameter. This would give us a model, M0 say, nested in M1 but with two fewer parameters. We need to be cautious about using this simple comparison for a variance parameter (see Chapter 1). Instead, we calculate the deviance under model M0; we call it $L0$. Its value is $L0 = 1424.31$. So the reduction in deviance on fitting the two extra parameters is $L0 - L1 = 17.30$, highly significant when tested as a value of χ^2 with two degrees of freedom ($p = 0.00018$). Clearly, model M0 is inadequate, and we stay with model M1.

The fixed coefficients of the variables x_{2jk} (income) and x_{3jk} (population density) do not reach significance, but we retain these two variables in the model in case they become informative in the context of any revised model. Our next step is to identify any statistically exceptional units (data points or

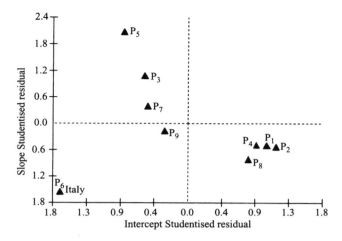

Figure 6.2   Studentised residuals for slopes versus intercepts at level 3 (nation).

explanatory points) or groups of points. We start at the highest level, countries (level 3), and focus on the relationship of mortality rate to the main explanatory variable ultraviolet index.

In the fitted relationship in (6.1), each of the nine countries has its own intercept $\beta_0 + v_{0k} = \beta_{0k}$, say, and slope β_{1k} with respect to ultraviolet index; for example country 9 (Netherlands) has intercept β_{09} and slope β_{19}. So the kth country ($k = 1, \ldots, 9$) has an intercept residual $\beta_{0k} - \beta_0 = v_{0k}$ and a slope residual $\beta_{1k} - \beta_1 = v_{1k}$. The MLwiN software gives us estimates $\hat{v}_{0k}$ and $\hat{v}_{1k}$ of these residuals and of their diagnostic standard errors (Goldstein, 1995, Appendix 2.2), which we denote by $se(\hat{v}_{0k})$ and $se(\hat{v}_{1k})$. We thus get the values of the Studentized residuals, $\hat{v}'_{0k}$ and $\hat{v}'_{1k}$ say, defined as

$$\hat{v}'_{0k} = \frac{\hat{v}_{0k}}{se(\hat{v}_{0k})}, \qquad \hat{v}'_{1k} = \frac{\hat{v}_{1k}}{se(\hat{v}_{1k})} \qquad (6.6)$$

In Figure 6.2, $\hat{v}'_{1k}$ is plotted against $\hat{v}'_{0k}$ for the nine countries ($k = 1, \ldots, 9$), giving points $P_1, \ldots, P_9$. If our assumption of (bivariate) normality is correct, the scatter of the points should be roughly according to a bivariate normal distribution with unit means and unit variances. In fact, the point P_6 is evidently outlying; this is the point for Italy.

6.4   USING DUMMY VARIABLES

To investigate this further, we introduce a dummy variable x_{4k}:

$$x_{4k} = \begin{cases} 1 & \text{for the units in country } k = 6 \text{ (Italy)}, \\ 0 & \text{for the units in the other 8 countries}, \end{cases} \qquad (6.7)$$

giving the augmented set of explanatory variables

$$x_0, x_{1ijk}, x_{2jk}, x_{3jk}, x_{4k}$$

and a model M2 given by (6.2) (6.3) and, instead of (6.1),

$$y_{ijk} = \beta_{0ijk}x_0 + \beta_{1k}x_{1ijk} + \beta_2 x_{2jk} + \beta_3 x_{3jk} + \beta_4 x_{4k}. \tag{6.8}$$

Fitting model M2 to the data gives a deviance $L2 = 1403.34$. The reduction $L1 - L2$ consequent on fitting one extra parameter is 3.67; testing this as a value of χ_1^2, its significance probability (p) is 0.055, indicating some evidence that the results for Italy are out of line with those for the other 8 countries. Pursuing this possibility, we examine the mean mortality rates for the 20 regions in Italy. In the map of Italy in Figure 6.3, the regions are shown with the mean mortality

Figure 6.3 Mean age-standardised mortality rates (per 100 000) for prostate cancer for Italian regions.

Table 6.3   Regions of Italy, with mean mortality rates (per 100 000) and number of areas (in paretheses) for each region.

Northern regions			Southern regions		
Valle d'Aosta	14.4	(1)	Lazio	9.9	(5)
Piemonte	13.4	(6)	Abruzzi	8.9	(4)
Lombardia	12.7	(9)	Molise	8.4	(2)
Trentino–Alto Adige	12.7	(2)	Campania	7.7	(5)
Veneto	11.4	(7)	Basilicata	8.3	(2)
Friuli–Venezia Giulia	14.3	(4)	Puglia	8.2	(5)
Liguria	12.7	(4)	Calabria	7.1	(3)
Emilia–Romagna	12.2	(8)	Sardegna	6.8	(4)
Toscana	10.8	(9)	Sicilia	7.3	(9)
Umbria	12.1	(2)			
Marche	10.8	(4)			

rate for the region marked in each region. It is evident that the country divides into 11 northern regions, with mean mortality rates from 10.8 to 14.4 per 100 000, and 9 southern regions, with mean mortality rates from 6.8 to 9.9 per 100 000; see Table 6.3.

This result is not surprising. In Italy, the southernmost of the countries in the survey, people tend to be genetically different in the south, and thus potentially less vulnerable to some of the carcinogenic effects of incident ultraviolet light. See Cavalli-Sforza *et al.* (1994).

To model this feature of the data, we replace the single country Italy at level 3 ($k = 6$) by two level-3 units, northern Italy ($k = n$, say) and southern Italy ($k = s$, say) as given by Table 6.3. We shall call these 'N. Italy' and 'S. Italy'; and, for ease of writing, we shall consider them as two separate 'countries', making now 10 countries in all.

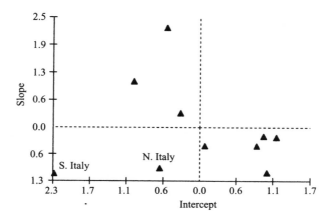

Figure 6.4   Studentised residuals for slopes versus intercepts at level 3 (nation), with Italy split into N. and S. Italy.

We first fit a model given by (6.1)–(6.3) to the data reclassified into 10 countries instead of 9. Figure 6.4 shows the plot of $\hat{v}'_{1k}$ against $\hat{v}'_{0k}$ analogously to Figure 6.2. This suggests that the results for N. Italy are in line with those for the eight non-Italian countries, whereas S. Italy is an outlying level-3 unit. Is it outlying in respect of the intercept β_{0s} or the slope β_{1s}, or both? To answer this, we model these as fixed parameters by introducing dummy variables x_{5k} and x_{6k} as follows:

$$x_{5k} = 1, \quad x_{6k} = x_{1ijs} \quad \text{for the units in country S. Italy,}$$
$$x_{5k} = 0, \quad x_{6k} = 0 \quad \text{for the units in the other 9 countries.}$$

We proceed to fit a model M3 given by (6.2), (6.3), and the following equation, with the set of explanatory variables $x_0, x_{1ijk}, x_{2jk}, x_{3jk}, x_{5k}, x_{6k}$:

$$y_{ijk} = \beta_{0ijk}x_0 + \beta_{1k}x_{1ijk} + \beta_2 x_{2jk} + \beta_3 x_{3jk} + \beta_5 x_{5k} + \beta_6 x_{6k}. \tag{6.9}$$

The estimates of the coefficients (fixed parameters) are

$$\left.\begin{array}{ll} \hat{\beta}_0 = 14.58\,(0.32), & \hat{\beta}_1 = -0.184\,(0.115), \\ \hat{\beta}_2 = -0.0009\,(0.0072), & \hat{\beta}_3 = 0.00019\,(0.00021), \\ \hat{\beta}_5 = -3.61\,(1.36), & \hat{\beta}_6 = -0.133\,(0.325), \end{array}\right\} \tag{6.10}$$

and the deviance $L3 = 1405.15$.

In (6.10), the estimate of β_5 differs significantly from zero, but that of β_6 does not. Accordingly we drop the dummy variable x_{6k} and adopt a model M4 given by (6.2), (6.3) and

$$y_{ijk} = \beta_{0ijk}x_0 + \beta_{1k}x_{1ijk} + \beta_2 x_{2jk} + \beta_3 x_{3jk} + \beta_5 x_{5k}. \tag{6.11}$$

The estimates of the fixed parameters under model M4 are

$$\left.\begin{array}{ll} \hat{\beta}_0 = 14.57\,(0.32), & \hat{\beta}_1 = -0.202\,(0.107), \\ \hat{\beta}_2 = -0.0007\,(0.0071), & \hat{\beta}_3 = 0.00018\,(0.00026), \\ \hat{\beta}_5 = -3.81\,(1.26). \end{array}\right\} \tag{6.12}$$

The estimates of the random parameters are

$$\left.\begin{array}{l} \text{level-3 var}\begin{pmatrix} v_{0k} \\ v_{1k} \end{pmatrix} = \begin{pmatrix} 0.249\,(0.306) \\ -0.023\,(0.082) & 0.077\,(0.046) \end{pmatrix}, \\ \text{level-2 var}(u_{0jk}) = 0.275\,(0.199), \\ \text{level-1 var}(e_{0ijk}) = 2.637\,(0.255). \end{array}\right\} \tag{6.13}$$

The deviance is $L4 = 1405.31$. $L4$ differs by an insignificant amount (0.16) from $L3$, confirming that the extra term $\beta_6 x_{6k}$ in model M3 is uninformative.

We have thus arrived at a model M4 that shows the responses (i.e. the mortality rates) in S. Italy to be systematically lower (by some 4 units) than the values that the data from the other countries would predict; but which shows the rate of change of response with ultraviolet index in S. Italy to be consistent with the rates of change observed in the other 9 countries.

6.5 LEVERAGE, DELETION RESIDUALS AND DFITS

Having dealt with the identification of statistically exceptional points at level 3, we move to the next level down: regions (level 2). A number of techniques are available; however, in view of various interrelationships existing between them, a few carefully chosen procedures can provide a good guide to exceptional points in the data, whether outliers, leverage points or influential points. Here, as in Langford and Lewis (1998), we choose to focus on Studentised residuals, on leverage values, and on a measure of influence called DFITS. We now give a brief account of these procedures. For further details of the diagnostic statistics involved and methods for their calculation, see Langford and Lewis (1998), Goldstein (1995), Hocking (1996) and Cook and Weisberg (1995).

Briefly recapitulating some standard results, we write a linear model as

$$E(\mathbf{y}) = \mathbf{X}\boldsymbol{\beta},$$

where $\mathbf{y}$ is the $n \times 1$ vector of observations (response variables), $\mathbf{X}$ is the $n \times q$ design matrix with q explanatory variables, and $\boldsymbol{\beta}$ is a $q \times 1$ vector of fixed coefficients. In an ordinary linear regression (i.e. single-level) model, the estimated (or fitted) response values $\hat{\mathbf{y}}$ are given by

$$\hat{\mathbf{y}} = \mathbf{X}\hat{\boldsymbol{\beta}} = \mathbf{H}\mathbf{y}, \tag{6.14}$$

where

$$\mathbf{H} = \mathbf{X}(\mathbf{X}^{\mathrm{T}}\mathbf{X})^{-1}\mathbf{X}^{\mathrm{T}} \tag{6.15}$$

which is the $n \times n$ projection or 'hat' matrix.

The n cases have n fitted residuals $(r_1, \ldots, r_n)$, given by the vector

$$\mathbf{r} = \mathbf{y} - \hat{\mathbf{y}} = (\mathbf{I}_n - \mathbf{H})\mathbf{y}. \tag{6.16}$$

In a multilevel model, with properties corresponding to (6.16), there will be several residuals, depending on the level. For a model with n_m units and q_m random coefficients at level m, there are $q_m n_m$ residuals at level m. We shall refer to the typical residual as r_{mcl}, the residual of the *c*th random coefficient $(c = 1, \ldots, q_m)$ for unit l $(l = 1, \ldots, n_m)$. Since diagnostic procedures are carried out at a particular level, we can omit the subscript m for convenience and write the residual as r_{cl}. Scaling r_{cl} by its 'diagnostic' standard error s_{cl} (see Goldstein, 1995, Chapter 2) we get the Studentised residual

$$r'_{cl} = r_{cl}/s_{cl}. \tag{6.17}$$

In our mortality data, for example, there were $n_3 = 9$ countries at level 3 and $q_3 = 2$ random coefficients (intercept and slope), giving the 2×9 Studentised residuals $\hat{v}'_{0k}$ and v'_{1k} $(k = 1, \ldots, 9)$ plotted in Figure 6.2 to identify possible outliers at country level. In our general notation, v'_{0k}, v'_{1k} are r'_{cl} with $c = 0, 1$ and $l = k$.

In calculating the standard error s_{cl} in (6.17), the data from all the n_m units are used. However, if unit l is outlying in respect of the random coefficient c, this will inflate the standard error and so deflate the Studentised residual r'_{cl}. To avoid this risk, we can calculate a fitted response value $\hat{y}_{l(l)}$ and a standard error $s_{cl(l)}$ from the data, omitting unit l. The Studentised residual r'_{cl} is then replaced as diagnostic by the deletion residual

$$r'_{cl(l)} = \frac{y_l - \hat{y}_{l(l)}}{s_{cl(l)}}. \tag{6.18}$$

Refitting of the model omitting unit l is not required, since $r'_{cl(l)}$ is directly related to r'_{cl}:

$$r'_{cl(l)} = r'_{cl} \left(\frac{n_m - q_m - 1}{n_m - q_m - r'^2_{cl}} \right)^{1/2}. \tag{6.19}$$

Other terminology is used by some authors. Our Studentised and deletion residuals are sometimes called internally Studentised and externally Student-ised, and sometimes called Standardised and Studentised. (We prefer to keep the term Standardised for scaling by a known (true) standard deviation and not an estimated one.) Our notation r_{cl} differs from the notations $\hat{p}_{hi}$ (which we would call $\mathbf{r}_{mc}$) in Goldstein (1995) and $\hat{p}_{mi}$ (likewise $\mathbf{r}_{mc}$) in Langford and Lewis (1998), which both do not use an index for the omitted unit (our l). We prefer to use r for a residual rather than p, in line with such texts as Hocking (1996), Cook and Weisberg (1982) and Chatterjee and Hadi (1988).

Studentised residuals, and more particularly deletion residuals, consistent with the model on which they are based, can be regarded as approximately normal $N(0,1)$ quantities. So, for diagnostic purposes, a value outside the range (say) -3 to $+3$ indicates that the data point is not consistent with the model. It is thus flagged as 'special'; the reason for this may usefully call for examination and interpretation.

Turning to leverage, the leverage for unit l at level m is based on h_{ll}, the lth diagonal element of the appropriate projection matrix $\mathbf{H}$ (see Langford and Lewis, 1998). This diagonal element is a quantity between 0 and 1; the nearer it is to 1, the greater is the distance between x_l, the explanatory variable value for unit l, and the other explanatory points. The trace of $\mathbf{H}$, $\sum_{l=1}^{n_m} h_{ll}$, is equal to q_m; accordingly, the measure of leverage we use in this chapter is the scaled quantity

$$h'_{ll} = h_{ll}/q_m. \tag{6.20}$$

Values of h'_{ll} greater than, say, $2.5/n_m$ may be regarded as exceptionally high, indicating that the inclusion of unit l in the data set substantially increases the precision of estimation of model parameters. We can call $2.5/n_m$ a 'calibration value' for this leverage diagnostic.

As a measure of influence we use the so-called DFITS (or DFFITS) statistic proposed by Belsley et al. (1980). The influence of unit l, i.e. the extent to which its inclusion in the data set changes the estimate of a model parameter, is clearly related to the effect on the prediction of case l caused by omission of that case, i.e. $\hat{y}_l - \hat{y}_{l(l)}$.

For diagnostic purposes, this is scaled by the estimated standard deviation of $\hat{y}_l$. We denote the residual mean square, again estimated from the data with case l omitted, by $s_{(l)}^2$. The true standard deviation of $\hat{y}_l$ is $\sigma\sqrt{h_{ll}}$, which we estimate by $s_{(l)}\sqrt{h_{ll}}$. Hence we get the DFITS measure of influence of unit l as

$$(DFITS)_l = \frac{\hat{y}_l - \hat{y}_{l(l)}}{s_{(l)}\sqrt{h_{ll}}} \,. \tag{6.21}$$

A suitable calibration value for this statistic is $2.5\sqrt{(q_m/n_m)}$; values of DFITS greater than this indicate high influence of case l, and the possible need to investigate this feature of the data.

We now resume the analysis of our mortality rate data, proceeding to level 2. Studentised residuals and deletion residuals for intercept, leverage values and values of DFITS are calculated for the 79 regions ($m = 2, n_2 = 79$). It is found that there are no unusual values for any of these statistics. This indicates that, on the assumption of model M4, there are no statistically exceptional regions. So, following our strategy of dealing with the levels in sequence from highest to lowest, we move on to level 1, areas ($m = 1, n_1 = 353$).

Figure 6.5 shows the 353 ordered values of Studentised residuals for intercept, plotted against their normal scores. There are two points outlying from the expected linear relationship; these correspond to the areas Alpes Maritimes in France (denoted by 'AM') and Longford in Ireland ('LO'). To remove possible bias, we calculate the deletion residuals; we find two outlying units AM and LO. Calculating the 353 leverage values, we find that there are two outlying points whose leverage values exceed a calibration level $2.5/n_1 = 2.5/353 = 0.0071$. These correspond to the areas Brussels ('BR') and Berlin West ('BW'). Figure 6.6 shows the leverage values plotted against the deletion residuals, with AM, LO, BR and BW appearing as outlying points on the scatter diagram.

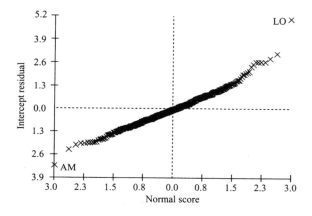

Figure 6.5   Level-1 standardised residuals versus normal scores.

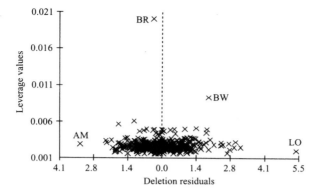

Figure 6.6   Leverage values versus deletion residuals (model with separate intercept for southern Italy).

BR and BW are areas whose mortality rates accord well with model M4, and are of high leverage; the data from these two areas are particularly useful in that they notably increase the precision of estimation of the modelled relationship.

Areas AM and LO are level-1 outliers relative to model M4. Are they also influential units? We have also calculated the distribution of the 353 values of DFITS. There is one area with a high DFITS value exceeding a calibration level $2.5\sqrt{(q_1/n_1)} = 2.5\sqrt{2/353} = 0.188$, and this is LO. So the level-1 outlier LO is influential, relative to model M4. In what respect is it influential? To answer this, we fit (temporarily) a model – call it M_{LO} – in which Longford is moved to the fixed part of the model, i.e. its coefficient is modelled as a fixed parameter. We introduce a dummy variable x_{7ijk} given by

$$x_{7ijk} = \begin{cases} 1 & \text{for Longford} \\ 0 & \text{for the other 352 areas.} \end{cases}$$

With the explanatory variables x_0, x_{1ijk}, x_{2jk} x_{3jk}, x_{5k}, x_{7ijk}, model M_{LO} is given by (6.2), (6.3), and

$$y_{ijk} = \beta_{0ijk}x_0 + \beta_{1k}x_{1ijk} + \beta_2 x_{2jk} + \beta_3 x_{3jk} + \beta_5 x_{5k} + \beta_7 x_{7ijk}. \tag{6.22}$$

On fitting model M_{LO}, the estimates of the coefficients (fixed parameters) are

$$\left.\begin{array}{ll} \hat{\beta}_0 = 14.55\,(0.33), & \hat{\beta}_1 = -0.195\,(0.107), \\ \hat{\beta}_2 = -0.0007\,(0.0073), & \hat{\beta}_3 = 0.00019\,(0.00026), \\ \hat{\beta}_5 = -3.86\,(1.29), & \hat{\beta}_7 = 9.05\,(1.57). \end{array}\right\} \tag{6.23}$$

The estimates of the random parameters are

$$\begin{aligned} \text{level-3 var}\begin{pmatrix} v_{0k} \\ v_{1k} \end{pmatrix} &= \begin{pmatrix} 0.285\,(0.336) & \\ -0.023\,(0.083) & 0.073\,(0.045) \end{pmatrix}, \\ \text{level-2 var}(u_{0jk}) &= 0.533\,(0.211), \\ \text{level-1 var}(e_{0ijk}) &= 2.208\,(0.220). \end{aligned} \tag{6.24}$$

The deviance $L_{LO} = 1375.31$. The reduction in deviance on fitting the extra parameter β_7 is $L4 - L_{LO} = 30.00$, which is highly significant as a value of χ^2 with 1 df ($p = 4 \times 10^{-8}$). This confirms the discordant status of the area LO. In plain terms, the mortality rate for Longford was much higher than what would have been expected from the main data set, the difference being some 9 units.

Comparing (6.23) and (6.24) with (6.12) and (6.13), the estimates of the fixed parameters other than β_7 are much the same under model M_{LO} as the corresponding values under model M4, and the same is true for the estimates of the random parameters at level 3. However, the estimate of the random parameter at level 2, $var(u_{0jk})$, is very different (0.533 as compared with 0.275). Here, then, is where LO is influential: its inclusion in the data set substantially decreases the estimate of the intercept variance at level 2.

This calls for comment. In a single-level regression, omission of an influential data point would normally decrease the variance. In our multilevel model, on the other hand, we have found that dummying out of an influential data point (Longford) at level 1 has actually increased the intercept variance at level 2. To see how this arises, we first note that the values of the ultraviolet index in Ireland were not typical of those for the other countries in the data set, being particularly low; the values of x_1 for the areas in Ireland ranged from -6.0 to -3.8, x_1 in our analysis being the value of the ultraviolet index centred on the mean for the entire data set.

There are four Irish regions, Connaught, Munster, Ulster and Leinster, and Longford is one of the 12 areas in Leinster. Figure 6.7 shows the scatter plot of y against x_1 for these 12 areas. Longford (LO) is clearly an outlier, as we already know. The fitted regression lines for all 12 points (solid line) and for the 11 points omitting LO (broken line) are also shown; their equations are

$$y = 7.59 - 1.50x_1 \quad \text{(including LO)}$$
$$y = 13.97 - 0.054x_1 \quad \text{(omitting LO)}.$$

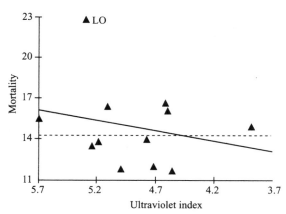

Figure 6.7   Scatter plot of age-standardised prostate cancer mortality (per 100 000) by ultraviolet index for Irish areas with regression lines including and excluding Longford (LO).

The mortality rate for LO was exceptionally high, whereas the intercept value 7.59 (Leinster including LO) is much lower than the value 13.87 (Leinster omitting LO). It is a matter of slope. The values of x_1 for the areas in Leinster range from -5.7 to -3.8; the value of x_1 at which the intercept is measured, namely $x_1 = 0$, lies well outside this range – in fact twice the width of the diagram frame to its right. The high y-value for the outlier LO in Figure 6.7 induces a negative regression slope (-1.50) and hence a low value (7.59) for the intersection of the regression line with $x_1 = 0$, i.e. the intercept. When LO is omitted, the regression line is practically horizontal and the intercept is consequently higher (13.97). This value is more consistent with the intercept values for the other Irish regions, but at the same time less consistent with those for the entirety of regions in the data set.

Essentially, then, the influential role of LO is the result of using ultraviolet index values which are not centred round the country mean (Ireland) but round the mean for the entire data set. If locally centred values had been used, LO would be the same outlier as before in the same data, but it would no longer be strongly influential. Thus we have arrived, perhaps unexpectedly, not at an intrinsic feature of Longford's prostate cancer mortality rate, but at a feature of the data-centring procedure.

The current model M4 must now be revised to remove the distorting effect of the two outlying level-1 units AM and LO. We already have a dummy variable x_{7ijk} for Longford, and now introduce an analogous dummy variable x_{8ijk} for Alpes Maritimes. With this additional explanatory variable, we set up a revised model M5 given by (6.2), (6.3) and

$$y_{ijk} = \beta_{0ijk}x_0 + \beta_{1k}x_{1ijk} + \beta_2 x_{2jk} + \beta_3 x_{3jk} + \beta_5 x_{5k} + \beta_7 x_{7ijk} + \beta_8 x_{8ijk}. \quad (6.25)$$

On fitting model M5, the estimates of the fixed parameters are

$$\left.\begin{aligned}
&\hat{\beta}_0 = 14.56\,(0.33), && \hat{\beta}_1 = -0.186\,(0.107), \\
&\hat{\beta}_2 = -0.0000\,(0.0072), && \hat{\beta}_3 = 0.00019\,(0.00025), \\
&\hat{\beta}_5 = -3.88\,(1.30), && \hat{\beta}_7 = 9.06\,(1.54), \\
&\hat{\beta}_8 = -5.52\,(1.60).
\end{aligned}\right\} \quad (6.26)$$

The estimates of the random parameters are

$$\left.\begin{aligned}
&\text{level-3 var}\begin{pmatrix} v_{0k} \\ v_{lk} \end{pmatrix} = \begin{pmatrix} 0.300\,(0.347) & \\ -0.015\,(0.084) & 0.072\,(0.045) \end{pmatrix}, \\
&\text{level-2 var}(u_{0jk}) = 0.525\,(0.205), \\
&\text{level-1 var}(e_{0ijk}) = 2.127\,(0.219).
\end{aligned}\right\} \quad (6.27)$$

The deviance is $L5 = 1363.59$. The reduction in deviance on fitting the extra parameter β_8 is $L_{LO} - L5 = 11.72$, which is highly significant as a value of χ^2 with 1 df ($p = 0.0006$), confirming the discordant status of area AM. The mortality rate for Alpes Maritimes was much lower than what would have been expected from the main data set, the difference being some 5 or 6 units.

Comparing (6.26), (6.23) and (6.12), the estimates of the fixed parameters other than β_8 under model M5 are little changed from their values under the previous models. Comparing (6.27) and (6.24), the same is true for the estimates of the random parameters under models M_{LO} and M5.

So, starting with model M1 as initial model, one cycle of identification of statistically exceptional points in the data has been completed with the setting up of model M5. We now iterate the modelling process, with model M5 as our new initial model. As a general procedure, this is necessary, because with a change of model a unit might change its status. For example, a non-outlying higher-level unit might become outlying relative to a revised model in which one of its component-lower level units has been effectively removed by modelling its coefficient as a fixed parameter. A leverage value or a DFITS value might also change substantially under a change of model.

As in the previous cycle, we find no unusual values for our chosen diagnostics at level 2, and so no evidence of any statistically exceptional regions. Proceeding to level 1, there are no further outlying areas. Relative to model M4, there were two high-leverage areas, BR and BW; after the separate modelling of LO and AM, the leverage values for BR and BW are no longer exceptional, and there are in fact no areas of high leverage relative to model M5. Area BW appeared quite influential under model M4, but its DFITS value under model M5 is much reduced and it is no longer influential.

We have thus reached a model M5 with all exceptional points in the data identified and accounted for, and with a satisfactory fit provided to all the rest of the data. No further iteration of the modelling process is called for.

6.6 CONCLUDING REMARKS

This chapter has been concerned with the identification of outliers, points of high leverage, and influential points in the analysis of multilevel data, and how to use them to assist in the interpretation of the data. We have presented three diagnostic tools for this purpose:

(i)   (internally) Studentised residuals, and deletion residuals, their externally Studentised equivalent;
(ii) leverage values;
(iii) DFITS values.

These have been used to reveal features of a three-level set of data on age-standardised mortality rates for cancer of the prostate, covering 353 different areas in nine European countries. We have found that the relationship of mortality rate to ultraviolet index can be regarded as having followed a common pattern for eight of the nine European countries and for northern Italy, but that throughout southern Italy the mortality rates for any given value of ultraviolet index were systematically lower than elsewhere by some 4 units. Among the areas, we have found that Alpes Maritimes in France had an unusually low mortality rate, while Longford in Ireland had an even more unusually high mortality rate.

These and other findings have been arrived at by use of the three diagnostic tools listed above. There are other diagnostic procedures in the literature, which we have not discussed. In the wise words of Cook and Weisberg (1995, pp. 8–9), 'If every recommended diagnostic is calculated for a single problem the resulting 'hodgepodge' of numbers and graphs may be more of a hindrance than a help and will undoubtedly take much time to comprehend. Life is short and we cannot spend an entire career on the analysis of a single set of data. The cautious analyst will select a few diagnostics for application in every problem . . .'

ACKNOWLEDGEMENTS

We are grateful to the Economic and Social Research Council for encouraging and supporting our research, by the award of Visiting Fellowships to one of us (TL) for the 1994–1995 session and to one of us (IHL) for the 1995–1996 session, to work with the Multilevel Models Project at the Institute of Education, University of London, in connection with the Analysis of Large and Complex Datasets Programme.

CHAPTER 7

Modelling Non-Hierarchical Structures

Jon Rasbash and William Browne

Mathematical Sciences, Institute of Education, University of London, UK

7.1 INTRODUCTION

In the models discussed in this book so far, we have assumed the populations from which data have been drawn are hierarchical. This assumption is not always justified. Two main types of non-hierarchical model are considered in this chapter: cross-classified models and multiple-membership models. This chapter draws on the work of Rasbash and Goldstein (1994) and Hill and Goldstein (1998).

7.2 CROSS-CLASSIFIED MODELS

This section is divided into five subsections. In Section 7.2.1, we look at situations in health research that can give rise to a two-way cross-classification and suggest some notation to describe this model. In Sections 7.2.2 and 7.2.3, we look at more complicated cross-classified structures and extend the notation. In Section 7.2.4, we describe general rules for the notation construction. In Section 7.2.5, we describe the analysis of an example data set.

7.2.1 Two-way cross-classifications – a basic model

Let us suppose that we have data on a large number of patients who attend many hospitals and that we also know the neighbourhood in which the patient lives; we regard patient, neighbourhood and hospital all as important sources of variation for the patient-level outcome measure we wish to study. Typically, hospitals will draw patients from many different neighbourhoods, and the

Multilevel Modelling of Health Statistics Edited by A.H. Leyland and H. Goldstein

Table 7.1   Patients cross-classified by hospital and neighbourhood.

	Neighbourhood 1	Neighbourhood 2	Neighbourhood 3
Hospital 1	×	× ×	
Hospital 2	× ×	× ×	×
Hospital 3	× × ×		× ×
Hospital 4	×	× ×	× ×
Hospital 5	× ×		

Table 7.2   Patients nested within hospitals nested within neighbourhoods.

	Neighbourhood 1	Neighbourhood 2	Neighbourhood 3
Hospital 1	× × ×		
Hospital 2		× × × × ×	
Hospital 3		× × × × ×	
Hospital 4	× × × × ×		
Hospital 5			× ×

inhabitants of a neighbourhood will go to many hospitals. No pure hierarchy can be found, and patients are said to be contained within a cross-classification of hospitals by neighbourhoods. This can be represented diagrammatically as in Table 7.1, for the case of twenty patients contained within a cross-classification of three neighbourhoods by five hospitals, with each patient represented by an ×.

In this example, we have patients at level 1, and neighbourhood and hospital are cross-classified at level 2. The characteristic pattern of a cross-classification is shown; some rows contain multiple entries *and* some columns contain multiple entries. In a nested relationship, if the row classification is nested within the column classification then all the entries across a row will fall under a single column, and vice versa if the column classification is nested within the row classification. For example, if hospitals are nested within neighbourhoods we might observe the pattern seen in Table 7.2.

We can consider other examples of two-way cross-classifications commonly occurring in health applications.

(i) Repeated measures contained within patients crossed by professionals   Here we have repeated measurements on patients, and at each occasion we have a response measurement taken by a professional, for example a nurse. Often a given patient will be measured by different professionals on different occasions. In this case, measurement occasions are contained within a cross-classification of individual by professional. This situation is shown in Table 7.3 for the case of three patients measured on three occasions and three nurses. Note that this table has the same formal structure as Table 7.1, with the measurement occasion this time being represented by an ×.

(ii) Health surveys with individuals grouped within interviewers by areas   It is often desirable in health surveys to separate individual, interviewer and area

Table 7.3 Measurement occasions cross-classified by patient and nurse.

	Nurse 1	Nurse 2	Nurse 3
Patient 1	×	××	
Patient 2	×	×	×
Patient 3	×		××

effects. When each interviewer works in multiple areas and each area is serviced by several interviewers, individuals are contained within a cross-classification of areas by interviewers.

(iii) Individuals within primary health care by secondary health care units Often we wish to know how much variability in a response is attributable to processes occurring at the primary health care level (e.g. general practitioners) and how much is attributable to processes occurring at the secondary health care level (e.g. hospitals). In this case, patients are contained within a cross-classification of GPs by hospitals.

More complex structures often arise, but before considering them, we set out the algebra for the basic two-way, normally distributed, cross-classified model with variance components. We can write this model as

$$\left. \begin{aligned} y_{i(j_1,j_2)} &= (X\beta)_{i(j_1,j_2)} + u_{j_1} + u_{j_2} + e_{i(j_1,j_2)}, \\ u_{j_1} &\sim N(0, \sigma_{u1}^2), \\ u_{j_2} &\sim N(0, \sigma_{u2}^2), \\ e_{i(j_1,j_2)} &\sim N(0, \sigma_e^2), \end{aligned} \right\} \tag{7.1}$$

where $(X\beta)_{i(j_1,j_2)}$ is the linear predictor. Using the example of measurement occasions within a cross-classification of patients by consultants, $y_{i(j_1,j_2)}$ is the response measurement at occasion i, contained in the cell defined by patient j_1 and consultant j_2, u_{j_1} is the random effect for patient j_1, u_{j_2} is the random effect for consultant j_2, and $e_{i(j_1,j_2)}$ is a level-1 residual for the ith repeated measurement contained in the cell defined by patient j_1 and consultant j_2.

7.2.2 More complex cross-classified population structures

In the previous example, with four consultants (c1–c4), four patients (p1–p4) and four measurement occasions per patient (×) a basic structure can be represented diagrammatically as shown in Table 7.4.

Table 7.4 Occasions cross-classified by patient and consultant.

	c1	c2	c3	c4
p1	××	×		×
p2	×		×	×
p3	×		××	×
p4		××		××

Table 7.5 Occasions cross-classified by patient and consultant, with patient and consultant jointly nested within hospital.

	c1		c2		c3		c4	
	h1	h2	h1	h2	h1	h2	h1	h2
p1	××		××					
p2	× × ×		×					
p3						××		××
p4						×		× × ×

If we introduce hospitals into the model, three possible structures arise. First, both patients and consultants may be nested within hospital, but patient is still crossed with consultant. If we introduce two hospitals, with patients 1 and 2 and consultants 1 and 2 in hospital 1 and patients 3 and 4 and consultants 3 and 4 in hospital 2, we have the structure illustrated in Table 7.5. Here level 1 is measurement occasion ($\times$), level 2 is a cross-classification of consultant (c1–c4) by patient (p1–p4) nested within level-3 units, hospitals (h1 and h2). We can write a variance components model as

$$
\left.
\begin{aligned}
y_{i(j_1,j_2)k} &= (X\beta)_{i(j_1,j_2)k} + v_k + u_{j_1 k} + u_{j_2 k} + e_{i(j_1,j_2)k}, \\
v_k &\sim N(0, \sigma_v^2), \\
u_{j_1 k} &\sim N(0, \sigma_{u1}^2), \\
u_{j_2 k} &\sim N(0, \sigma_{u2}^2), \\
e_{i(j_1,j_2)k} &\sim N(0, \sigma_e^2).
\end{aligned}
\right\}
\qquad (7.2)
$$

Here $y_{i(j_1,j_2)k}$ is the response measure taken on the ith measurement occasion contained in the cell defined by patient j_1 and consultant j_2 within hospital k. The term $u_{j_1 k}$ is the random effect for the j_1th patient in hospital k, $u_{j_2 k}$ is the random effect for the j_2th consultant in hospital k, and $e_{i(j_1,j_2)k}$ is the level-1 residual for the ith repeated measurement contained in the cell defined by patient j_1 and consultant j_2 in hospital k.

The second possible structure is where patients are nested within hospitals, but consultants work in more than one hospital. In this case, we have the structure indicated in Table 7.6. Here occasions ($\times$) are at level 1 and patients (p1–p4) at level 2 are nested within hospitals (h1 and h2) at level 3. Consultant

Table 7.6 Occasions cross-classified by consultants and (patient within hospital).

	c1		c2		c3		c4	
	h1	h2	h1	h2	h1	h2	h1	h2
p1	×				×		×	
p2	×				××			
p3				×		×		×
p4		×				×		×

Table 7.7 Occasions cross-classified by patients, consultants and hospitals.

	c1		c2		c3		c4	
	h1	h2	h1	h2	h1	h2	h1	h2
p1	×	×						×
p2		××			×		×	
p3	×					×	×	
p4	×			×				

(c1–c4) is crossed with the patient-within-hospital nested structure, and is also defined at level 3. Note that if consultants are crossed at level 3 with hospital, they are assumed to be crossed with any units nested within hospitals. We can write this model as

$$
\left.
\begin{aligned}
y_{i(j_1 k_1, k_2)} &= (X\beta)_{i(j_1 k_1, k_2)} + v_{k_1} + v_{k_2} + u_{j_1 k_1} + e_{i(j_1 k_1, k_2)}, \\
v_{k_1} &\sim N(0, \sigma_{v_1}^2), \\
v_{k_2} &\sim N(0, \sigma_{v_2}^2), \\
u_{j_1 k_1} &\sim N(0, \sigma_u^2), \\
e_{i(j_1 k_1, k_2)} &\sim N(0, \sigma_e^2).
\end{aligned}
\right\}
\qquad (7.3)
$$

Here i indexes occasion, j_1 indexes patients, who are nested within hospitals, k_1 indexes hospitals, and k_2 indexes consultants. The subscript notation $i(j_1 k_1, k_2)$ refers to the ith measurement occasion for patient j_1 nested within hospital k_1, where the measurement was made by consultant k_2.

Finally, patients may be seen by different consultants on different occasions, consultants work in more than one hospital and patients move between hospitals. Here, measurement occasions ($\times$) are at level 1 contained within a three-way cross-classification of patient (p1–p4) by consultant (c1–c4) by hospital (h1 and h2) at level 2. This situation is represented by Table 7.7. We can write this model as

$$
\begin{aligned}
y_{i(j_1, j_2, j_3)} &= X\beta + u_{j_1} + u_{j_2} + u_{j_3} + e_{i(j_1, j_2, j_3)}, \\
u_{j_1} &\sim N(0, \sigma_{u1}^2), \\
u_{j_2} &\sim N(0, \sigma_{u2}^2), \\
u_{j_3} &\sim N(0, \sigma_{u3}^2), \\
e_{i(j_1, j_2, j_3)} &\sim N(0, \sigma_e^2),
\end{aligned}
\qquad (7.4)
$$

Here i indexes measurement occasion, and j_1, j_2 and j_3 index patient, hospital and consultant respectively.

7.2.3 Models where more than one cross-classified structure exists

In the examples given so far, we have basic or 'atomic' units that lie within a cross-classification defined by higher-level units. For example, with measurement occasions contained within a cross-classification of patient by consultant, measurement occasions are the atomic units, and patient and consultant are the higher-level units defining the cross-classification. It is possible for units (e.g. patients) to be both a higher-level cross-classifying factor and also an atomic unit within a further higher-level cross-classification. Consider the case where

(i)   repeated measures are contained within a cross-classification of patients by consultants;
(ii) patients are themselves contained within a cross-classification of general practitioners by hospitals;
(iii) consultants work in many hospitals and see patients from many general practitioners.

We have two separate cross-classified structures operating here. We can represent them both schematically. The first involves repeated measurement occasions, patients and consultants, and can be drawn as shown in Figure 7.1, in which the atomic units are repeated measures occasions. The second cross-classified structure, shown in Figure 7.2, involves patients as the atomic units contained within a cross-classification of general practitioner by hospital. Both structures can be included in a single diagram, as shown in Figure 7.3. The indexing notation for the complete structure is

$$y_{i(j(k_1, k_2), k_3)},$$

where $i(j(k_1, k_2), k_3)$ refers to the ith measurement on the jth patient, referred by general practitioner k_1 to hospital k_2, and the measurement was made by consultant k_3.

7.2.4   General rules for notation construction

There are as many subscript indices as there are classifications present. In Figure 7.3, we have five classifications – occasion, patient, GP, hospital and consultant – and therefore five subscript indices: i, j, k_1, k_2 and k_3. There are as many numerical sub-subscripts as there are hierarchies present in the data. In Figure 7.3, we have three hierarchies present:

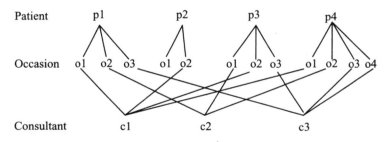

Figure 7.1   Occasion within patient by consultant structure.

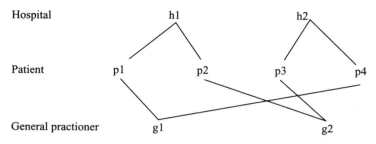

Figure 7.2 Patient within GP by hospital structure.

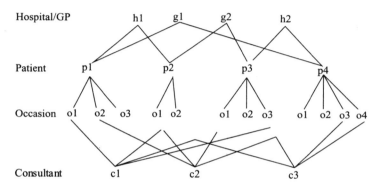

Figure 7.3 Combined structure for occasion within patient by consultant structure and patients within GP by hospital.

(i)   occasion i nested within patient j nested within GP k_1;
(ii)   occasion i nested within patient j nested within hospital k_2;
(iii) occasion i nested within consultant k_3.

Therefore numerical sub-subscripts run from 1 to 3. If a subscript index has no numerical subscript, this means that the classification it refers to belongs to more than one hierarchy. Hence i has no numerical subscript, because occasions belong to all three hierarchies. Also, j has no numerical subscript, because the patient belongs to both the hospital and GP hierarchies.

Subscript indices separated by commas and enclosed in parentheses specify a cross-classified relationship. To the left of the parentheses are the lower-level units that are contained within the cross-classification. Any list of indices not separated by commas specifies a purely nested relationship. We can parse the overall indexing structure $y_{i(j(k_1,k_2),k_3)}$ to produce separate indexing structures for each hierarchy present:

y_{ijk_1} occasion within patient within general practitioner;
y_{ijk_2} occasion within patient within hospital;
y_{ik_3} occasion within consultant.

If the deepest level of nesting in any of the hierarchies present is two then the greatest (alphabetical) subscript present will be *j;* if the deepest level of nesting present is three then the greatest (alphabetical) subscript present will be *k*; and so on. The crossed hierarchies present may contain different levels of nesting. In Figure 7.3, we have two hierarchies with three levels of nesting, and one hierarchy with two levels of nesting. The top level in each crossed hierarchy is given the same subscript letter, which is defined by the deepest level of nesting present; these subscripts are differentiated by the hierarchy number sub-subscript. Within each hierarchy, any nested levels unique to that hierarchy are subscripted down from the top subscript letter, and any nested levels common to more than one hierarchy are subscripted up from *i*. Thus, if we had two hierarchies present,

occasion : patient : consultant : ward : hospital,
occasion : patient : GP : GP practice,

we should write $y_{ij(k_1 l_1 m_1, l_2 m_2)}$, where i, j, k_1, l_1, m_1, l_2 and m_2 represent occasion, patient, consultant, ward, hospital, GP and GP practice respectively.

7.2.5   An example analysis for a cross-classified model

We consider a data set concerning artificial insemination by donor. Detailed descriptions of this data set and the substantive research questions addressed by modelling it within a cross-classified framework are given in Echoard and Clayton (1998). The data were re-analysed in Clayton and Rasbash (1999) as an example case study demonstrating the properties of a data augmentation algorithm for estimating cross-classified models.

The data consist of 1901 women who were inseminated by sperm donations from 279 donors. Each donor made multiple donations; there were 1328 donations in all. A single donation is used for multiple inseminations. An attempt by a woman to conceive involves a series of monthly inseminations, with one insemination per ovulatory cycle. An attempt is ended when a woman successfully conceives or after a period of one year if no conception occurs. After a first success – the conception and birth of a child – a second series of inseminations can be attempted. In some cases, several such series have been attempted by the same woman. The data contain 12 100 cycles within 2437 attempts within the 1901 women.

There are two crossed hierarchies: a hierarchy for donors and a hierarchy for women. Level 1 corresponds to measures made at each ovulatory cycle. The response that we analyse is the binary variable indicating if conception occurs in a given cycle. The hierarchy for women is cycles within attempts within women. The hierarchy for donors is cycles within donations within donors. Within a series of cycles, a women may receive sperm from multiple donors/ donations. The model is represented schematically in Figure 7.4. Here cycles are positioned on the diagram within attempts within women, so the topology of the diagram reflects the hierarchy for women. When we connect the male hierarchy to the diagram, we see crossing connections between donations and cycles, revealing the crossed structure of the data set.

Women

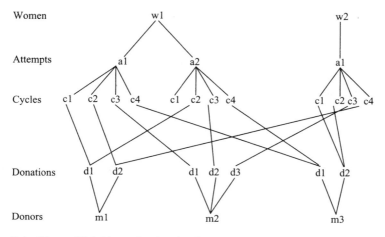

Figure 7.4 The artificial insemination by donor example.

We can write the model as

$$
\begin{aligned}
y_{i(j_1 k_1, j_2 k_2)} &\sim \mathrm{Bin}(1, \pi_{i(j_1 k_1, j_2 k_2)}) \\
\mathrm{logit}(\pi_{i(j_1 k_1, j_2 k_2)}) &= (X\beta)_{i(j_1 k_1, j_2 k_2)} + v_{k_1} + u_{j_1 k_1} + v_{k_2} + u_{j_2 k_2}, \\
v_{k_1} &\sim N(0, \sigma_{v1}^2), \qquad v_{k_2} \sim N(0, \sigma_{v2}^2), \\
u_{j_1 k_1} &\sim N(0, \sigma_{u1}^2), \qquad u_{j_2 k_2} \sim N(0, \sigma_{u2}^2),
\end{aligned}
\right\} \tag{7.5}
$$

where i indexes cycle, j_1 indexes attempts, k_1 indexes women, j_2 indexes donations and k_2 indexes donors. The binary response $y_{i(j_1 k_1, j_2 k_2)}$ for the ith cycle from the j_1th attempt from the k_1th women, that received sperm form the j_2th donation from the k_2th donor, is assumed to follow a binomial (Bernoulli)

Table 7.8 Results for the artificial insemination data.

Parameter	Description	Estimate	SE
β_0	Intercept	−4.04	2.30
β_1	Azoospermia[a]	0.22	0.11
β_2	Semen quality	0.19	0.03
β_3	Woman's age > 35	−0.30	0.14
β_4	Sperm count	0.20	0.07
β_5	Sperm motility	0.02	0.06
β_6	Insemination to early	−0.72	0.19
β_7	Insemination to late	−0.27	0.10
σ_{v1}^2	Variance between women	1.02	0.21
σ_{u1}^2	Variance between attempts	0.644	0.21
σ_{v2}^2	Variance between donors	0.338	0.07
σ_{u2}^2	Variance between donations	0.131	0.068

[a] Fecundability of women not impaired.

distribution (see Chapter 3). The attempts, women, donations and donor variance components are σ_{u1}^2, σ_{v1}^2, σ_{u2}^2 and σ_{v2}^2 respectively. The results of fitting this model are shown in Table 7.8.

After inclusion of covariates, there is considerably more variation in the probability of a successful insemination attributable to the women hierarchy than the donor hierarchy.

7.3 MULTIPLE-MEMBERSHIP MODELS

When lower-level units are influenced by more than one higher-level unit from the same classification, we have a multiple-membership model. For example, if patients are treated by several nurses then patients are said to be 'multiple members' of nurses. Each of the nurses treating a patient contributes to the treatment outcome. Following Hill and Goldstein (1998), with a slight change in notation, a two-level multiple membership model can be written as

$$\left.\begin{aligned}
y_{i\{j\}} &= (X\beta)_{i\{j\}} + \sum_{h \in \{j\}} u_h p_{ih} + e_{i\{j\}}, \\
u_h &\sim N(0, \sigma_u^2), \\
e_{i\{j\}} &\sim N(0, \sigma_e^2), \\
\sum_h \pi_{ih} &= 1 \quad \forall i,
\end{aligned}\right\} \tag{7.6}$$

where $\{j\}$ is the full set of level-2 units (nurses in the above example). The level-1 units (patients) are indexed uniquely by i and may be a 'member' of more than one nurse. The index h uniquely indexes nurses, and p_{ih} is a predetermined weight declaring the proportion of membership of patient i to nurse h. For example, if we knew that a quarter of patient i's treatment was administered by nurse h then a weight of 0.25 might be reasonable. Often we will not have information at this level of detail, in which case we may assume equal weights.

To clarify this, consider a simple example. Suppose we have four patients treated by up to two of three nurses (n1, n2, n3). The weighted membership matrix P might look like that shown in Table 7.9. Here patient 1 was seen by nurse 1 and 3 but not nurse 2, patient 2 was only seen by nurse 1, and so on. If we substitute the values of p_{ih}, i and h from this table into (7.6), we get the following series of equations:

Table 7.9 An example weighted membership matrix for patients and nurses.

	n1 ($h=1$)	n2 ($h=2$)	n3 ($h=3$)
p1 ($i=1$)	0.5	0	0.5
p2 ($i=2$)	1	0	0
p3 ($i=3$)	0	0.5	0.5
p4 ($i=4$)	0.5	0.5	0

$$
\left.
\begin{aligned}
y_{1\{j\}} &= (X\beta)_{1\{j\}} + 0.5u_1 + 0.5u_3 + e_{1\{j\}}, \\
y_{2\{j\}} &= (X\beta)_{2\{j\}} + 1u_1 + e_{2\{j\}}, \\
y_{3\{j\}} &= (X\beta)_{3\{j\}} + 0.5u_2 + 0.5u_3 + e_{3\{j\}}, \\
y_{4\{j\}} &= (X\beta)_{4\{j\}} + 0.5u_1 + 0.5u_2 + e_{4\{j\}}.
\end{aligned}
\right\}
\tag{7.7}
$$

7.3.1 An example analysis for a multiple-membership model

The example considered now is not medical but from the related field of veterinary epidemiology. The data have kindly been supplied by Mariann Chriel. They are concerned with causes and sources of variability in outbreaks of salmonella in flocks of chickens in poultry farms in Denmark between 1995 and 1997. The data have a complex structure; there are two main hierarchies. The first is concerned with production: level 1 units are flocks of chickens and the response is binary, indicating whether or not there was any instance of salmonella in the flock. Flocks live for a short time, about two months, before they are slaughtered for consumption. Flocks are kept in houses, so in a year a house may have a throughput of five or six flocks. Houses are grouped together in farms. The data comprise 10 127 child flocks in 725 houses in 304 farms.

The second hierarchy is concerned with breeding. There are two hundred parent flocks in Denmark, and eggs are taken from the parent flocks to four hatcheries. After hatching, the chicks are transported to the farms in the production hierarchy, where they form the production (child) flocks. Any given child flock may draw chicks from up to six parent flocks. Child flocks are therefore multiple members of parent flocks. Since chicks from a single parent flock go to many production farms, and chicks on a single production farm come from more than one parent flock, this means that the multiple-membership breeding hierarchy is cross-classified with the production hierarchy. We can represent this situation as in Figure 7.5, which shows a similar

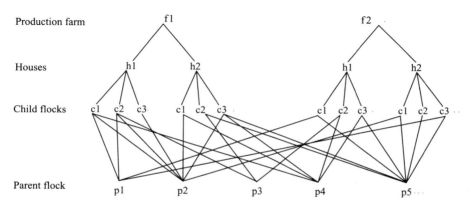

Figure 7.5 The Danish poultry example with multiple membership and a cross-classification.

pattern to Figure 7.4; the crossing appears in the diagram when we connect child flock to parent flock. However, unlike Figure 7.4, where each cycle is connected to only one donation, here each child flock is connected to multiple parent flocks.

The epidemiological questions of interest revolve around how much of the variation of salmonella incidence is attributable to each of houses, farms and parent flocks. The notation construction rules for crossed hierarchies where hierarchies contain multiple membership relations is as described in Section 7.2.4, except that where a lower-level unit (child) is a multiple member of a higher-level unit (parent), the higher-level index is enclosed in braces. Therefore, for the salmonella data, we can write the model as

$$
\left.
\begin{aligned}
& y_{i(\{k_1\}, j_2 k_2)} \sim \text{Bin}(1, \pi_{i(\{k_1\}, j_2 k_2)}), \\
& \text{logit}(\pi_{i(\{k_1\}, j_2 k_2)}) = (X\beta)_{i(\{k_1\}, j_2 k_2)} + \sum_{h \in \{k_1\}} v_{k_1} p_{i k_1} + v_{k_2} + u_{j_2 k_2}, \\
& v_{k_1} \sim N(0, \sigma_{v1}^2), \qquad v_{k_2} \sim N(0, \sigma_{v2}^2) \\
& u_{j_2 k_2} \sim N(0, \sigma_u^2),
\end{aligned}
\right\}
\quad (7.8)
$$

where i, j and k_2 uniquely index child flocks, houses and production farms, and k_1 is the full set of parent flocks. Parent flocks are defined at level 3, and hence are denoted by k_1, because parent flocks are cross-classified with production farms. Therefore the cross-classification is at level 3, and parent flocks are automatically cross-classified with houses, which are nested within production farms.

Five covariates were considered in the model: year = 1996, year = 1997, hatchery = 2, hatchery = 3, and hatchery = 4. The intercept therefore corresponds to hatchery 1 in 1995. The results are shown in Table 7.10.

There is a large parent flock variance, indicating that a parent flock process is having a substantial effect on the variability in the probability of child flock infection. This could be due to genetic variation across parent flocks in resistance to salmonella, or it may due to differential hygiene standards in parent flocks resulting in certain parent flocks introducing infected eggs into the

Table 7.10   Results for Danish poultry data.

Parameter	Description	Estimate	SE
β_0	Intercept	−1.86	0.187
β_1	1996	−1.04	0.131
β_2	1997	−0.89	0.151
β_3	Hatchery 2	−1.47	0.22
β_4	Hatchery 3	−0.17	0.21
β_5	Hatchery 4	−0.92	0.29
σ_{v1}^2	Variance between parent flocks	1.02	0.22
σ_{v2}^2	Variance between farms	0.59	0.11
σ_u^2	Variance between houses	0.19	0.09

system. There is also substantial variation across production farms and rather less between houses within farms.

7.4 SOFTWARE AND COMPUTATIONAL ISSUES

The software packages HLM (Bryk *et al.*, 1996), SAS (SAS, 1992), BUGS (Spiegelhalter *et al.*, 1995) and MLwiN (Rasbash *et al.*, 1999a,b) are all capable of fitting non-nested random effects models. The various methods used to estimate non-nested random effects models by these packages are computationally intense – either in terms of storage requirements or CPU time or, in some cases, both. The details of cross-classified model specification and estimation in the IGLS/RIGLS framework are set out in Rasbash and Goldstein (1994). The multiple-membership model is set out in Goldstein and Hill (1998), with the reasons for its computational intensity given in Bull *et al.* (1999). An alternative block-sampling MCMC approach for estimating cross-classified models, which promises to solve many of the computational problems, is set out in Clayton and Rasbash (1999).

7.5 CONSEQUENCES OF IGNORING NON-HIERARCHICAL STRUCTURES

Analysing only hierarchical components of populations that have additional non-nested structures has two potentially negative consequences. First, the model is under-specified because there are sources of variation that have not been included in the model. This under-specification can lead to an under-estimation of the standard errors of the parameters and therefore to incorrect inferences. Secondly, the estimated variance components obtained from the simple hierarchical model, or sets of separate hierarchical models, are not reliable. They may change substantially if the additional non-nested structures are included in a single model. For example, we may wish to know about the relative importance of general practices and hospitals on the variation in a patient-level outcome. If patients are cross-classified by hospital and general practice, we need to fit the full cross-classified model including patients, general practices and hospitals in order to address this question. Looking at two separate hierarchical analyses – one of patients within hospital, the other of patients within general practices – is not sufficient.

The examples in this chapter have dealt with variance components models. Extensions to random slopes models for cross-classified and multiple-member structures are readily available.

CHAPTER 8

Multinomial Regression

Min Yang

Mathematical Sciences, Institute of Education, University of London, UK

8.1 INTRODUCTION

In the fields of medicine and public health, many outcomes are discrete categories: for example, the health status of individuals in relation to their daily life style of smoking, drinking and exercise habits, attitudes of individuals towards health services, and measures assessing the effects of some public health education programs. In some cases, outcomes may be recorded on a nominal scale: for example, healthy, having an infectious disease and having a non-infectious disease. These categories are independent of each other. In other cases, they may be ordinal, say very satisfactory, fairly satisfactory and not satisfactory, or rarely ill, often ill, severely ill and hospitalised during a given period of life. In such studies, hierarchical structures or clustering effects often manifest themselves owing to the common practice of multiple-stage sampling involved in social surveys (see Chapter 11). Alternatively, the hierarchy may be defined by the geographical nature of the social structure: individuals may be nested within health authorities, or patients nested within clinics and clinics nested within health authorities. In modelling this kind of data, both the clustering effect and the discrete error distribution of the outcomes should be taken into account.

In this chapter, we introduce multilevel models for multiple categorical responses with multinomial error distributions for both nominal and ordinal data. Goldstein (1987, 1995) proposed fitting such models using IGLS (iterative generalised least squares) in a multilevel framework. The models in this chapter have been fitted in the multilevel modelling software MLwiN (Rasbash *et al.* etc 1999a,b) using macros written for this purpose (Yang *et al.*, 1998; see also Chapter 13). We shall use two examples from public health to illustrate these models, their underlying assumptions and the interpretation of parameter estimates.

Multilevel Modelling of Health Statistics Edited by A.H. Leyland and H. Goldstein
©2001 John Wiley & Sons, Ltd

8.2   MODELLING NOMINAL RESPONSES WITH A THREE-LEVEL HIERARCHY

8.2.1   Example: Use and abuse of antibiotics for acute respiratory tract infections in young children (Li 1997; Kunin *et al.*, 1990)

Acute respiratory tract infection (ARI) is a prevalent disease in childhood. As a consequence of inadequate treatment for ARI due to either failed diagnosis, or drug resistance of the bacteria, or inappropriate self-medication, pneumonia is a leading cause of death in young children in the rural communities of many developing countries. In the late 1980s, the World Health Organisation (WHO) introduced a program of case management for ARI in young children under five years old, with the aim of reducing the mortality caused by ARI. The program was implemented in some rural communities in China in the early 1990s (WHO, 1991a,b; Li, 1997).

Antibiotics were very easy to obtain in China for three reasons. First, modern and advanced medical equipment was beyond the financial resources of many of the low-level clinics. Medical diagnosis was mainly reached through some basic laboratory tests and the doctor's clinical experiences. Secondly, a large proportion of practitioners in rural areas did not hold qualifications from medical colleges but qualified through special training courses. They lacked the knowledge, experience and equipment to make a correct diagnosis in many cases. Hence many resorted to using single or combined antibiotics for every possible infectious case, no matter whether they were necessary or not, i.e. a so-called 'shotgun' strategy inevitably became common practice in disease treatment. Finally, many antibiotics were on sale to the public in any pharmacist's shop. One of the consequences of this antibiotic abuse that has caused the most concern in the context of ARI-related mortality is drug resistance.

To analyse the pattern of doctors' prescribing behaviour and the associated factors in antibiotic abuse in some areas of rural China, a total of 135 doctors in 36 hospitals in two counties of China were investigated regarding their prescribing practice. One county was in the WHO program, the other was not. Both counties had similar capital income per head. There were a total of 855 treated cases of ARI in children under five. Their medical records were checked for symptoms and clinical signs, the medicine prescribed, and the correct final diagnosis obtained. Their parents were asked whether they had administered any self-medication before seeing their doctors and the reasons for it. Based on the WHO case management criteria, the doctor's antibiotic prescription in each case was classified into three categories: (1) combined antibiotic abuse without clinical indicators, (2) abuse of one antibiotic without clinical indicators, and (3) correct use with clinical indicators. The percentages of the categories in the two counties are given in Table 8.1. This shows that doctors abused antibiotics in over half of the ARI cases overall, and there was some difference in the doctors' prescription of antibiotics between the two counties. Other variables available from this study are shown in Table 8.2.

Table 8.1 Number (percentage) of cases in the three categories of prescribing behaviour.

County	Category 1	Category 2	Category 3	Total
A (in the program)	64 (51.2)	50 (40.0)	11 (8.9)	125 (100.0)
B (not in the program)	197 (27.0)	330 (45.2)	203 (27.8)	730 (100.0)
TOTAL	261 (30.5)	380 (44.4)	214 (25.0)	855 (100.0)

Table 8.2 Explanatory variables and their codes/values.

	Variable	Description	Code/value
Patient level	x_1	Age group in years	0–5
	x_2	Temperature	36.0–40.4 °C, centred at 36 °C
	x_3	Free medicine	0 = yes, 1 = no
	x_4	Self medication	0 = no, 1 = yes
	x_5	Correct diagnosis	0 = yes, 1 = no
Doctor level	x_6	Education level	Six categories from medical school to self-taught
	x_7	Length of practice in years	1–45, centred around 25
Hospital level	x_8	Location	1 = county, 2 = town, 3 = village
	x_9	WHO ARI program	0 = yes, 1 = no

The doctor's prescribing behaviour is a rather complicated issue. Many social and psychological factors from both the patient's and the doctor's sides can be influential. Culture, education and policy can also be associated with this behaviour. This small study aims only to look into some basic issues and to provide some possible evidence for future studies. The basic issues are how the patient's status on arrival at the clinic may affect the doctor's prescribing behaviour, how the doctor's own experience may affect prescribing, and what the effect of the hospital's location is. Whether or not the area is participating in the WHO program is only used as another characteristic of hospital location.

8.2.2 The basic model for nominal responses with three-level hierarchies

For the ith patient treated by the jth doctor in the kth hospital, the probability of the patient being in one of the doctor's prescription groups is $\pi_{ijk}^{(h)}, h = 1, 2, 3$, where $\sum_h \pi_{ijk}^{(h)} = 1$. Treating the third category of correct use of antibiotics with clinical indicators as the base group t and taking into account the clustering of patients nested within doctor within hospital, we have the simplest three-level multinomial logistic regression model without fitting any of the explanatory variables given in Table 8.2:

$$\log\left(\frac{\pi_{ijk}^{(s)}}{\pi_{ijk}^{(t)}}\right) = \beta_0^{(s)} + v_{0k}^{(s)} + u_{0jk}^{(s)}, \qquad s = 1, \cdots, t-1. \tag{8.1}$$

Here $\pi_{ijk}^{(t)}$ is the probability for the ith patient in the base category, and $\beta_0^{(s)}$ is the logarithm of the ratio of the overall probability of a patient being in category s to the base group. A positive estimate of $\beta_0^{(s)}$ suggests a larger probability of being in category 1 or 2 than in the base category 3. The term $v_{0k}^{(s)}$ is the residual of the kth hospital's log ratio from the overall ratio β_0 for the corresponding category and has zero mean and variance $\sigma_{v0}^{2(s)}$. In other words, the log ratio for the kth hospital is $\beta_{0k}^{(s)} = \beta_0^{(s)} + v_{0k}^{(s)}$. Similarly, the term $u_{0jk}^{(s)}$ to be estimated is the residual of the jth doctor's log ratio from the hospital log ratio $\beta_{0k}^{(s)}$ for the corresponding category and has zero mean and variance $\sigma_{u0}^{2(s)}$. The log ratio for the jth doctor is $\beta_{0jk}^{(s)} = \beta_0^{(s)} + v_{0k}^{(s)} + u_{0jk}^{(s)}$. With the two residual variances, also known as random effects at levels 3 and 2, we can estimate the coverage of the log ratio among doctors as well as among hospitals. Any significant change in the random effects due to adding or removing a variable from the model indicates the degree to which that variable accounts for the contextual effects at different levels.

With only three categories in total ($s = 1$ for category 1, $s = 2$ for category 2 and $t = 3$ for category 3), (8.1) presents two models for the log ratios $\log(\pi_{ijk}^{(1)}/\pi_{ijk}^{(3)})$ and $\log(\pi_{ijk}^{(2)}/\pi_{ijk}^{(3)})$ respectively for binary data. For multiple categories, we fit (8.1) as a multivariate model (see Chapter 5). That is, for each patient, we have a vector of responses consisting of 0 and 1 according to the prescribing category. The ratio of the probability of being in any one category to that of the base category is then fitted in one model:

$$\log\left(\frac{\pi_{ijk}^{(s=1,2)}}{\pi_{ijk}^{(t=3)}}\right) = (\beta_0^{(1)} + v_{0k}^{(1)} + u_{0jk}^{(1)})z_{ijk} + (\beta_0^{(2)} + v_{0k}^{(2)} + u_{0jk}^{(2)})(1 - z_{ijk}). \tag{8.2}$$

The indicator variable z takes the value 1 for category 1, and 0 otherwise. The main effects of any explanatory variables are fitted by creating interactions between the variable and the indicator z and its complement $(1 - z)$.

The variance structure at level 3 (the hospital level) is given by

$$\text{var}(v_{0k}^{(1)} + v_{0k}^{(2)}) = \sigma_{v0}^{2(1)} + 2\sigma_{v0}^{(1,2)} + \sigma_{v0}^{2(2)}.$$

The covariance between categories 1 and 2 of the hospital log ratios is estimated. A positive estimate implies that a hospital with a higher proportion of category-1 prescriptions tends to have a higher proportion of category-2 prescriptions. The correlation coefficient can also be calculated as $\sigma_{v0}^{(1,2)}/(\sigma_{v0}^{(1)}\sigma_{v0}^{(2)})$.

In the same manner, the variance structure at level 2 (the doctor level) is as follows:

$$\text{var}(u_{0jk}^{(1)} + u_{0jk}^{(2)}) = \sigma_{u0}^{2(1)} + 2\sigma_{u0}^{(1,2)} + \sigma_{u0}^{2(2)}.$$

In addition to the fixed effects of the explanatory variables, it is of interest to explore whether the random variation at the hospital and doctor levels can be explained by the hospital and doctor variables in this study.

Taking the anti-logit of model (8.2), for patients in categories 1 and 2, we have the probability

$$\pi_{ijk}^{(s)} = \exp(\beta_0^{(s)} + v_{0k}^{(s)} + u_{0jk}^{(s)})[1 + \sum_{h=1}^{t-1} \exp(\beta_0^{(h)} + v_{0k}^{(h)} + u_{0jk}^{(h)})]^{-1}, \tag{8.3}$$

and for a patient in category 3, it is $1 - \sum_s \pi_{ijk}^{(s)}$.

Because for each patient there is a vector of $(0,1)$ responses Y_{ijk} according to the patient's category, it is natural to assume a multinomial error distribution to model (8.2) with parameter vector π_{ijk}. The error structure for the response vector is

$$\text{cov}(y_{ijk}^{(s)}, y_{ijk}^{(s')}) = \begin{cases} \pi_{ijk}^{(s')}(1 - \pi_{ijk}^{(s)})/n_{ijk} & (s' = s), \\ -\pi_{ijk}^{(s')}\pi_{ijk}^{(s)}/n_{ijk} & (s' \neq s), \end{cases} \tag{8.4}$$

where the denominator n_{ijk} is a constant (equal to one) in this case. Conditional on the other parameter estimates of the model, this covariance structure can be imposed at the patient level (Goldstein, 1995). On the diagonal of the covariance matrix (i.e. where $s' = s$), the response in each category has a binomial variance. Off the diagonal ($s' \neq s$), we have a covariance between categories.

To assess whether the model distribution assumption is adequate in these data, we can fit extra-multinomial variation parameters in (8.4). (See Chapter 3 for a discussion of extra-binomial variation.) To do so, we rewrite (8.4) as

$$\text{cov}(y_{ijk}^{(s)}, y_{ijk}^{(s')}) = \begin{cases} \pi_{ijk}^{(s')}(w_1 - w_2\pi_{ijk}^{(s)})/n_{ijk} & (s' = s), \\ -w_2\pi_{ijk}^{(s')}\pi_{ijk}^{(s)}/n_{ijk} & (s' \neq s), \end{cases} \tag{8.5}$$

where w_1 and w_2 are terms to be estimated at the patient level and are constrained to be equal. We can either set the constraint $w_1 = w_2 = 1$, assuming a multinomial error distribution for the data, or set the constraint that $w_1 = w_2 = w$, where w is freely estimated. If it is estimated to be close to 1, the model is adequate. If it is smaller or larger than one according to the significance level of the approximate Wald test (Goldstein, 1995), there is evidence of extra-multinomial variation. However, a simulation study (Yang, 1997) showed that even the best available estimation procedure (penalised quasi-likelihood, PQL, with a second-order approximation in MLwiN; Goldstein and Rasbash, 1996) resulted in an under-dispersion of about 10%. This was shown for both nominal and ordinal responses, whilst the estimates for other fixed and random effects in the model were not significantly different. In this case, we shall assume a multinomial distribution and constrain those terms to be equal to one.

Another model choice is to use the log–log link function for the probability of not being in category s:

$$\log[-\log(1 - \pi_{ijk}^{(s=1,\,2)})] = (\beta_0^{(1)} + v_{0k}^{(1)} + u_{0jk}^{(1)})z_{ijk} + (\beta_0^{(2)} + v_{0k}^{(2)} + u_{0jk}^{(2)})(1 - z_{ijk}).$$

The interpretation of the parameters here is different from when using the logit function, although other assumptions about the random effects and the multinomial distribution remain the same, and in most cases the two will give the similar results.

8.2.3 Fitting fixed effects

The explanatory variables can be fitted in a straightforward manner in model (8.2). To examine the effect of the doctor-level variable x_5, how the correct diagnosis affects the doctor's prescription of antibiotics, we fit model (8.6) with the same random effects at levels 2 and 3:

$$\log\left(\frac{\pi_{ijk}^{(s=1,\,2)}}{\pi_{ijk}^{(t=3)}}\right) = (\beta_0^{(1)} + \beta_5^{(1)}x_{5,\,jk} + v_{0k}^{(1)} + u_{0jk}^{(1)})z_{ijk} \tag{8.6}$$
$$+ (\beta_0^{(2)} + \beta_5^{(2)}x_{5,\,jk} + v_{0k}^{(2)} + u_{0jk}^{(2)})(1 - z_{ijk}).$$

The estimate of $\beta_5^{(1)}$ is interpreted as the log odds-ratio of incorrect diagnosis over correct diagnosis for the combined antibiotic abuse category, i.e.

$$\beta_5^{(1)} = \log\left(\frac{\pi_{\text{incorr}}^{(1)}\big/\pi_{\text{incorr}}^{(3)}}{\pi_{\text{corr}}^{(1)}\big/\pi_{\text{corr}}^{(3)}}\right).$$

The odds ratio of the incorrect diagnosis over the correct one is estimated by $e^{\beta_5^{(1)}}$. Similarly $\beta_5^{(2)}$ is the log odds-ratio of incorrect diagnosis over correct diagnosis for the category of single antibiotic abuse. Positive estimates suggest that a doctor who could not make a correct diagnosis when seeing a patient at an early stage is more likely to use one or more antibiotics without sufficient clinical evidence. The 'shotgun' strategy seems to be in operation. The difference between the estimates of $\beta_5^{(1)}$ and $\beta_5^{(2)}$ can be tested using an approximate Wald test.

For a model with a log–log link function, the estimate of $\beta_5^{(1)}$ is expressed as a log ratio between the correct diagnosis group and the incorrect diagnosis group, i.e.

$$\log\left[\frac{-\log(1 - \pi_{ijk}^{(1)})_{\text{incorr}}}{-\log(1 - \pi_{ijk}^{(1)})_{\text{corr}}}\right].$$

Taking the exponent $e^{\beta_5^{(1)}}$ gives us the form $\log(1 - \pi_{ijk}^{(1)})_{\text{incorr}}\big/\log(1 - \pi_{ijk}^{(1)})_{\text{corr}}$. We can see that a large value of $\beta_5^{(1)}$ suggests a *lower* chance of *not being* (or a *greater* chance of *being*) in the category of combined antibiotic abuse for a doctor who has incorrectly diagnosed his or her patients compared with one giving a correct diagnosis.

In the same manner, the effects of other explanatory variables are additive in model (8.6). For a continuous variable such as the patient's temperature x_2, the corresponding effect on any category $\beta_2^{(1)}$ or $\beta_2^{(2)}$ is interpreted as the change in the log ratio of combined or single antibiotic abuse over correct use as the patient's temperature increases by one degree.

Comparing the random effects at hospital and doctor levels between models (8.2) and (8.6) enables us to attribute the variation in prescribing behaviour between doctors or between hospitals according to the doctors' clinical diagnosis of the patients. We shall illustrate the analysis in the following subsection.

8.2.4   Fitting fixed effects in the example data

Table 8.3 presents estimates for four models. Model A is a null model fitting just the means for each category. Model B includes patient-level fixed effects, model C includes doctor-level characteristics, and finally hospital-level variables are fitted in model D.

In model A, we can observe the random effects for categories 1 and 2 among doctors with little covariance between the two categories. At the hospital level, the greatest variation between hospitals in terms of the probability of abusing antibiotics occurs for category 1, combined antibiotic abuse, with an estimated variance of 2.213. The variation for category 2 is not significant ($\chi_1^2 = 2.64, p = 0.104$). A small covariance at this level may imply that hospitals with greater probability of combined antibiotic abuse are also likely to have a greater probability of single antibiotic abuse. However, with no significant variation in single antibiotic abuse, this small covariance can also be ignored. Fitting an extra-multinomial variance parameter at level 1 gives us an estimate of 0.803 (SE = 0.030); this under-estimation occurs despite the use of second- order PQL (Goldstein and Rasbash, 1996) for estimation to overcome the bias associated with the 1st order and MQL procedure (Breslow and Clayton, 1993). Since this is only an unconditional variance components model and, based on the findings of the simulation study (Yang, 1997), we can leave the model as it is for further analysis.

In model B, we see that all the patient variables are significantly associated with the doctor's prescription of antibiotics. The fixed effects can be interpreted as follows. With increasing age, a child is more likely to be the victim of antibiotic abuse. As his temperature increases, he is much less likely to be such a victim. If the patient does not receive free medicine, the doctor is more likely to prescribe inappropriately, with the estimated odds-ratios for categories 1 and 2 being $e^{1.288} = 3.63$ (95% confidence interval 2.01–6.55) and $e^{1.011} = 2.75$ (1.64–4.61) respectively. A patient who has taken some medication before seeing the doctor is less likely to receive prescriptions inappropriately, with estimated odds-ratios of 0.60 (0.38–0.92) for category 1 and 0.81 (0.54–1.20) for category 2. The most significant factor is whether the doctor gave a correct diagnosis at an early consultation. The odds-ratio for reaching an incorrect diagnosis over the correct one is estimated to be 11.13 (7.22–11.17) for single antibiotic abuse and 6.69 (4.21–10.65) for combined abuse.

Table 8.3 Parameter estimates from fitting (8.2) and (8.6).

Parameter	Category	Model A	Model B	Model C	Model D
Fixed					
β_0	1	−0.604 (0.287)	−0.686 (0.335)	−3.344 (0.703)	−5.160 (0.927)
	2	0.416 (0.140)	0.284 (0.282)	−0.701 (0.484)	−0.288 (0.554)
β_1	1		0.223 (0.070)**	0.213 (0.071)**	0.020 (0.069)
	2		0.316 (0.066)**	0.283 (0.066)**	0.133 (0.064)*
β_2	1		−0.679 (0.109)**	−0.743 (0.110)**	−0.670 (0.115)**
	2		−1.325 (0.111)**	−1.221 (0.109)**	−1.186 (0.110)**
β_3	1		1.288 (0.302)**	1.743 (0.331)**	2.199 (0.407)**
	2		1.011 (0.264)**	0.925 (0.274)**	1.424 (0.303)**
β_4	1		−0.518 (0.223)*	−0.913 (0.245)**	−1.074 (0.246)**
	2		−0.216 (0.202)	−0.479 (0.211)*	−0.535 (0.207)*
β_5	1		1.901 (0.237)**	2.051 (0.242)**	2.178 (0.248)**
	2		2.410 (0.221)**	2.359 (0.223)**	2.410 (0.219)**
β_6	1			−0.135 (1.211)	0.060 (1.096)
				2.639 (0.632)**	1.619 (0.700)*
				3.012 (0.744)**	1.242 (0.779)
				3.287 (0.677)**	1.551 (0.826)
				4.070 (1.137)**	2.085 (1.274)
	2			0.298 (0.631)	−0.191 (0.675)
				0.919 (0.414)*	0.112 (0.454)
				0.868 (0.541)	−0.579 (0.656)
				0.627 (0.455)	−0.929 (0.599)
				1.951 (0.998)*	−0.021 (1.110)
β_7	1			0.013 (0.013)	—
	2			−0.029 (0.012)*	—
β_8	1 Town				1.645 (0.582)**
	Village				1.510 (0.372)**
	2 Town				2.610 (0.701)**
	Village				2.191 (0.524)**
β_9	1				1.771 (0.531)**
	2				0.693 (0.343)*
Random	Level 3				
$\sigma_{v0}^{2(1)}$		2.213 (0.706)	0.758 (0.294)	0.337 (0.203)	0.515 (0.262)
$\sigma_{v0}^{(1,2)}$		0.551 (0.257)	−0.120 (0.160)	0.146 (0.114)	0.074 (0.134)
$\sigma_{v0}^{2(2)}$		0.234 (0.168)	0.144 (0.139)	0.051 (0.118)	0.038 (0.118)
	Level 2				
$\sigma_{u0}^{2(1)}$		0.523 (0.227)	0.556 (0.216)	0.658 (0.244)	0.718 (0.247)
$\sigma_{u0}^{(1,2)}$		−0.080 (0.167)	−0.556 (0.185)	−0.650 (0.201)	−0.765 (0.210)
$\sigma_{u0}^{2(2)}$		0.715 (0.207)	0.573 (0.214)	0.611 (0.220)	0.678 (0.223)
	Extra variation		1.050 (0.040)	1.062 (0.040)	

*$p < 0.05$; ** $p < 0.01$.

It is interesting that patients administering self-medication and those in receipt of free medicine are less likely to be inappropriately prescribed antibiotics under this model. In reality, patients receiving free medicine were frequently from state-run institutions or enterprises. Their family incomes were reasonably stable, the parents had better education, and they received better state health service. It was easy for them to obtain surplus antibiotics for their first-aid box at home, and to self-prescribe them whenever they felt the need. In these data, 79.6% of cases receiving free medicine had self-administered antibiotics before seeing a doctor. As a consequence of self-prescribing, the illness of the patients may not appear so serious in the doctor's eyes or the infection may already be under control to a certain extent. On the other hand, patients not in receipt of free medicine were usually of a low socio-economic level or from the countryside where accessing health services was difficult. They might not see a doctor unless the illness was severe or became very complicated. This would be more likely to lead to the doctor applying a shotgun strategy in order to reduce the infection at the first visit.

Studying the random effects at the hospital level, the most dramatic reduction seems to be the variance of the combined antibiotic abuse, $\sigma_{v0}^{2(1)}$, compared with model A. The estimation procedure for model B was second-order marginal quasi-likelihood (MQL), since the preferred PQL procedure did not converge using MLwiN. The parameter estimates, especially the random effects, could therefore be under-estimated in model B. The different estimation procedures mean that we should not make the comparison between models A and B. The random effects at the doctor level are still moderate, suggesting that the patient variables do not account for the variation in outcomes at the doctor level. At hospital level, the variance for category 1 is still significant ($\chi_1^2 = 6.66, p < 0.05$).

In model C, the two doctor variables – clinical experience in years and professional education (ranging from medical college qualification to self-taught 'witch doctor'-type practitioners on an ordinal six-point scale) – are added in. The estimates of doctor's experience show a small effect on the category-2 outcome; the more clinical experience a doctor has, the less likely he is to commit single antibiotic abuse. As far as education is concerned, all of the estimates but one have positive signs. The effects for category 1 are more significant than they are for category 2. The general suggestion is that the lower the level of professional training the doctor received, the more likely he or she is to abuse antibiotics. Comparing the random effects between models B and C (which are estimated under the same procedure), we also find that the two doctor-level variables have explained more than half of the hospital-level variation. For example, the estimated variance of a category-1 outcome (combined abuse) is 0.758 in B and 0.337 in C, a reduction of 56%. In fact, none of the three terms is significantly greater than zero using the Wald test. This implies that the apparent differences between hospitals in the doctors' prescribing behaviour is mainly due to the variability of the doctors' professional experience and training.

In model D, we finally add in the two hospital-level variables: hospital location (town vs county, village vs county) and whether the hospital is in the

WHO program for ARI case management. The variable doctor's experience is left out of this model to avoid numerical problems. (The parameter estimate associated with this variable becomes insignificant when included in a model including hospital variables.) Table 8.3 shows that hospitals *not* in the WHO program are about 2.5 times more likely than hospitals *in* the program to abuse multiple antibiotics than they are to abuse single antibiotics, with the associated odds-ratio for this category being 5.87 (2.08–16.62). As there are no baseline data from the two counties, we can only presume that the difference between them may be caused partly by the impact of the WHO program and partly by their initial difference.

The most important factor at hospital level associated with the absence of antibiotic abuse is the hospital's location or authority level. Based on the estimates in Table 8.3, a hospital in a town is 8.94 times more likely than a county hospital to abuse single antibiotics, with 95% confidence interval (CI) 3.20–24.98, and 4.5 times more likely to abuse multiple antibiotics, with 95% CI 2.19–9.35. Similarly, a village hospital is 13.6 (1.24–53.73) times more likely than a county hospital to abuse single antibiotics and 5.18 (1.66–16.21) times more likely to abuse multiple antibiotics. This is because in China the public health service has a social hierarchy of county–town–village, with a large degree of variability in terms of the equipment and the manpower resources as well as differences between the patients themselves at each layer of the structure. The lower the level at which a hospital lies within the structure, the poorer its resources, and the poorer the patients it receives. From the data we can classify the 134 doctors by their qualifications or professional training and the hospital level they work for. We get the following picture:

	With qualification	With no qualification
County	33 (40.7%)	3 (5.7%)
Town	43 (53.1%)	9 (17.0%)
Village	5 (6.2%)	41 (77.4%)

This also explains why most effects of the doctor's education in model D have become insignificant compared with model C.

Apart from the patient's age, all other patient-level variables remain significant in this model.

8.2.5 Examining the residuals

We have routinely assumed in the models considered to date that the residuals at levels 2 (doctor) and 3 (hospital) are multivariate normal. As a diagnostic tool, it is always useful to examine the estimated higher-level residuals to make sure that the assumptions we placed on the model are valid.

In MLwiN, shrunken residuals or posterior means conditional on the fixed effect estimates as well as the random effect estimates are calculated (Goldstein, 1995). Based on model D in Table 8.3, the normal plots for the standardised level-3 residuals of the log odds for each outcome category are as shown in Figure 8.1. They show reasonable marginal normal distributions.

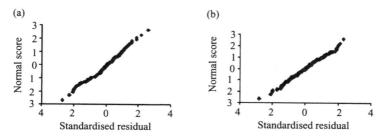

Figure 8.1 Normal plots for the residuals for (a) single antibiotic and (b) combined antibiotic abuse.

Another use of residual examination is to pinpoint units with very large or small residual estimates. From the modelling point of view, they may be outliers and need to be treated separately from other units (Langford and Lewis, 1998; see also Chapter 6). From a more substantive point of view, they are indicators identifying those units with discrepant outcomes that may need to be investigated further. The latter issue has been widely discussed and applied in the fields of health and education (Goldstein and Spiegelhalter, 1996; see also Chapter 9). However, we shall not deal with this issue in this chapter.

8.2.6 Fitting a random coefficient model

Having fitted model D, in which the main effects of all explanatory variables were estimated, we can further question the variability of the effects of any of the variables across hospitals or between doctors. For example, we can ask whether the effect of a doctor's incorrect diagnosis varies between doctors. This means that we are allowing the estimate of $\beta_5^{(s)}$ to vary across doctors, i.e. the estimated effect of the outcome for each category for each doctor is $\beta_{5jk}^{(s)} = \beta_5^{(s)} + u_{5jk}^{(s)}$, which has a mean of $\beta_5^{(s)}$ and variance $\sigma_{u5}^{2(s)}$. Substituting $\beta_{5jk}^{(s)} - u_{5jk}^{(s)}$ for $\beta_5^{(s)}$ in (8.6), we specify a more complex covariance structure at the doctor level with estimates:

	$u_{0jk}^{(1)}$	$u_{0jk}^{(2)}$	$u_{5jk}^{(1)}$	$u_{5jk}^{(2)}$				
$u_{0jk}^{(1)}$	$\sigma_{u0}^{2(1)}$				1.91 (0.74)			
$u_{0jk}^{(2)}$	$\sigma_{u0}^{(1,2)}$	$\sigma_{u0}^{2(2)}$			= −0.73 (0.21)	1.37 (0.54)		
$u_{5jk}^{(1)}$	$\sigma_{u05}^{(1)}$	×	$\sigma_{u5}^{2(1)}$		−1.17 (0.73)		1.01 (0.82)	
$u_{5jk}^{(2)}$	×	$\sigma_{u05}^{(2)}$	×	$\sigma_{u5}^{2(2)}$	−0.63 (0.53)			0.57 (0.62)

We have left out three covariance elements in the structure, since we consider them to have no substantive interpretation in this example. The four terms on the diagonal are the random effects of the two 'intercepts' associated with the base groups and the random effects of the two 'slopes' associated with the variable x_5.

This model gives us the distribution of the effect of variable x_5 among doctors as $\hat{\beta}_5^{(1)} \pm \sqrt{1.01}$ for combined abuse and $\hat{\beta}_5^{(2)} \pm \sqrt{0.57}$ for single abuse. However, the random effects for x_5 are not significant according to the Wald test. The covariance between the intercept and the slope for combined antibiotic abuse is -1.17, and the estimate for single antibiotic abuse is -0.63. Again, none of these is significantly different from zero.

We have allowed the coefficients of other patient level variables to vary among doctors at level 2. The only significant coefficients are $\beta_2^{(1)}$ associated with the patient's measured temperature, x_2 for combined antibiotic abuse and $\beta_2^{(2)}$ for single antibiotic abuse. The estimate of the fixed effect $\beta_2^{(1)}$ is $-0.661(\pm 0.129)$, and the variance between doctors is then related to the patient's temperature x_2 according to the function $1.317 - 2 \times 0.529x_2 + 0.329x_2^2$. The estimate of the fixed effect $\beta_2^{(3)}$ is $-1.149(\pm 0.117)$ and the corresponding variance between doctors is $(1.775 - 2 \times 0.50x_2 + 0.192x_2^2)$. Note that both variances are quadratic functions of the patient's temperature. It is also worth remarking that in this analysis the variable relating to the doctor's education has been removed from the fitted model.

8.3 MODELLING MULTILEVEL ORDINAL RESPONSES

8.3.1 Example: Television School and Family Smoking Prevention and Cessation Project (TVSFP)

This example is a subset from the TVSFP study, which was carried out during 1986–1988 (Flay *et al.*, 1989, 1995) and which was designed to test independent and combined effects of a school-based social-resistance curriculum and a television-based programme in terms of tobacco use prevention and cessation. The original design of the study forms a 2×2 classification of social-resistance classroom curriculum (CC or x_2 : yes = 1 and no = 0) by mass-media intervention (TV or x_3: yes = 1 and no = 0). CC and TV are the two major explanatory variables. The outcome, a tobacco and health knowledge scale (THKS) after intervention, is scored ordinally from 0 to 7, but has been regrouped into four categories (with 1 = 0–1, 2 = 2, 3 = 3 and 4 = 4–7). The baseline variable preintervention PRETHKS (x_1) is scored from 0 to 7, and is treated as a continuous variable. The study involved students from Los Angeles and San Diego. Randomisation to various design conditions was at the school level, while much of the intervention was delivered to students within classrooms.

This subset of the data consists of 1600 students from 135 classrooms in 28 schools. It is unbalanced, with a range of 1–13 classrooms per school and 2–28 students per classroom. The raw frequency distribution of the outcome is in Table 8.4. The cumulative proportion is our response.

Hedeker and Gibbons (1994) fitted 2–level ordinal regression models (ignoring classrooms) to the data using the maximum marginal likelihood estimation procedure in the MIXOR package. In this section, we illustrate three-level multinomial models for ordinal outcomes using the iterative generalised least

Table 8.4 Distribution of the Tobacco and Health Knowledge Scale (THKS).

Scale (s)	Frequency	Proportions	Cumulative proportion
1	355	0.222 (π_1)	0.222 (γ_1)
2	398	0.249 (π_2)	0.471 (γ_2)
3	400	0.250 (π_3)	0.721 (γ_3)
4	447	0.279 (π_4)	1.000 (γ_4)
TOTAL	1600	1.000	

squares estimation procedure in the MLwiN program and compare the results between the two procedures.

8.3.2 The basic model for ordinal responses with three-level hierarchies

For the ith student in the jth classroom of the kth school, there is a vector of responses taking the value 0 or 1 according to into which category the ordinal score falls, $(y_{ijk}^{(1)} y_{ijk}^{(2)} \ldots y_{ijk}^{(t)})$. In this example, $t = 4$. For a student in the first category, the associated response vector is (1 1 1 1). For a student in the second category, the response vector is (0 1 1 1), and that for a student in the third category is (0 0 1 1). The last category is always one for all students, forming the upper limit boundary of the responses. Leaving out the last category, we define the element of the vector as

$$E\left(y_{ijk}^{(s)}\right) = \gamma_{ijk}^{(s)} = \sum_{h=1}^{s} \pi_{ijk}^{(h)}, \qquad s = 1, \ldots, t-1, \qquad (8.7)$$

where $y_{ijk}^{(s)}$ are the observed cumulative proportions out of a total n_{ijk} (which is always one in our case) and s indexes the ordered cumulative categories. If we assume an underlying multinomial distribution for the category probabilities, the cumulative proportions have a covariance matrix given by

$$\text{cov}\left(y_{ijk}^{(s)}, y_{ijk}^{(s')}\right) = \frac{w\gamma_{ijk}^{(s)}\left(1 - \gamma_{ijk}^{(s')}\right)}{n_{ijk}}, \qquad s \le s'. \qquad (8.8)$$

The term w is constrained to be one if the variances are assumed to be strictly multinomial. In the same way as when fitting a nominal multinomial model, the extra-distributional variance can be tested by estimating w and its standard error freely within students. The Wald test can be used to test the significance of the estimate.

The most popular model to fit ordinal data, including the fixed effects of three explanatory variables as well as the interaction between x_1 and x_2, is written as

$$\log\left(\frac{\gamma_{ijk}^{(s)}}{1 - \gamma_{ijk}^{(s)}}\right) = \alpha^{(s)} - \left(\sum_{l=1}^{4} \beta_l x_{lijk} + v_k + u_{jk}\right). \qquad (8.9)$$

This model is also called the proportional odds model, and does not change the ordinal property of the raw scale using a logistic link, since

$$
\gamma_{ijk}^{(s)} = \left(1 + \exp\left\{ -\left[\alpha^{(s)} - \left(\sum_{l=1}^{4} \beta_l x_{lijk} + v_k + u_{jk} \right) \right] \right\} \right)^{-1}, \qquad (8.10)
$$

where the series of $\alpha^{(s)}$ account for the order of the proportional odds or the cumulative probabilities with the requirement that $\alpha^{(1)} \leq \alpha^{(2)} \dots \leq \alpha^{(t-1)}$. This requirement ensures that the larger the proportion $\gamma_{ijk}^{(s)}$ is, the greater log odds it has. The anti-logit value of the $\alpha^{(s)}$ can be interpreted as the threshold of the probability for a subject appearing in the category $s - 1$ before appearing in category s. The negative sign of the parameter β_1 means that we expect an increase over and above the $\alpha^{(s)}$ estimates for a positive effect of x_1. For a positive value of β_1, we expect a decrease in $\alpha^{(s)}$, and hence a negative effect on $\gamma_{ijk}^{(s)}$.

In the model, v_k is the random residual term at school level, assumed to be a normal variable from the distribution $N(0, \sigma_v^2)$. The model will estimate the single variance as the random effect of the overall proportional odds across schools accounting for the heterogeneity of schools. In other words, this model does not assume separate random effects for each of the $\alpha^{(s)}$ estimates. However, the model can be modified by assuming $v_k^{(s)} \sim N(0, \Omega_v)$ where Ω_v is a 3×3 covariance matrix for the random effects of the $\alpha^{(s)}$ at school level. We fit this model in Section 8.3.4.

Similarly, we have the same assumptions for and treatment of the residual term u_{jk} in the model at classroom level.

8.3.3 Modelling the example data

We first fit the model without including any explanatory variables as is shown in column A of Table 8.5. The estimate of the extra multinomial variation is 0.965, with a standard error of 0.020. We therefore have some confidence in our assumption that the response comes from a multinomial distribution.

In columns A–E of Table 8.5, we fit the fixed effects of the explanatory variables one at a time. From column B, we can see that all three estimates of the $\alpha^{(s)}$ become much larger than those in the null model in column A, and the students' knowledge of tobacco and health before the intervention is significantly associated with the outcome ($\chi_1^2 = 107.8$). The estimate 0.405 can be interpreted as the average increase in the proportional odds, which is accounted for by the pre-score measure. Based on the model described in (8.10), we obtain predictions of the $\alpha^{(s)}$ from the estimates given in columns A–C, and these are plotted in Figure 8.2. This shows that as prior knowledge of the issue increases by one unit on average, the expected increase in the outcome score is about 15% in category 1, 20% in category 2 and 13% in category 3.

The class-curriculum based programme estimated in column C shows a significant fixed effect ($\chi_1^2 = 21.98$). The prediction of the $\alpha^{(s)}$ in Figure 8.2 suggests that in addition to the prior knowledge effect, being in the

Table 8.5 Parameter estimates from fitting the three-level proportional odds models to the TVSFP data.

Parameter	Model A	Model B	Model C	Model D	Model E
Fixed effect					
$\alpha^{(1)}$	−1.398	−0.623	−0.228	−0.199	−0.106
	(0.120)	(0.134)	(0.135)	(0.155)	(0.171)
$\alpha^{(2)}$	−0.127	0.693	1.078	1.107	1.198
	(0.114)	(0.134)	(0.136)	(0.157)	(0.172)
$\alpha^{(3)}$	1.067	1.940	2.315	2.345	2.435
	(0.117)	(0.142)	(0.146)	(0.165)	(0.181)
β_1, Prethks		0.405	0.415	0.415	0.415
		(0.039)	(0.039)	(0.039)	(0.039)
β_2, CC			0.708	0.708	0.899
			(0.151)	(0.150)	(0.213)
β_3, TV				0.056	0.239
				(0.150)	(0.207)
β_4, CC*TV					−0.373
					(0.296)
Random effect					
σ_v^2	0.222	0.188	0.049	0.048	0.047
	(0.092)	(0.082)	(0.042)	(0.042)	(0.041)
σ_u^2	0.193	0.179	0.172	0.171	0.161
	(0.066)	(0.065)	(0.063)	(0.062)	(0.061)

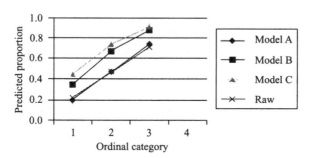

Figure 8.2 Predicted $\alpha^{(s)}$ based on the estimates of models A–C in Table 8.5.

class-curriculum based programme has an additional effect of 9.4% for category 1, 7.9% for category 2 and 3.6% for category 3.

It is worth noting that this variable alone has remarkably reduced the school-level random effect from 0.188 in model B to 0.049 in model C (i.e. by 73.9%), so that the school-level variation is small enough to be negligible. It is of interest to know whether this accountability of the CC program for the school level variation applies to all schools, or only to a few extreme schools. The residual estimates $\hat{v}_k$ and the standardised residuals for schools are plotted in

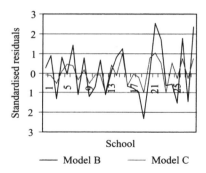

Figure 8.3 Plots of standardised school residuals for models B and C, with the fixed effect of the CC programme included in model C.

Figure 8.3, which shows that the residuals of all schools are reduced considerably when the CC programme fixed effect is included. This implies that the CC programme not only improves the students' knowledge of tobacco and health on average, but also accounts for over 70% of the variation in outcomes across schools.

From the estimates in columns D and E of Table 8.5, we find no significant fixed effect from the TV or the mass-media intervention programme ($\chi_1^2 = 0.14$), nor of an interaction effect between the CC and TV intervention programmes ($\chi_1^2 = 1.59$).

8.3.4   Fitting a random coefficient model

As far as the class-level random effect is concerned, its estimate is consistent for all models from A (0.193) to E (0.161) in Table 8.5. None of the explanatory variables has explained the random effect to any significant degree. It may be the case that there are no class-level variables that can explain the differences between classes, nor do the student level variables such as socio-economic factors reduce the heterogeneity within classes in this data set. What we can explore is allowing the fixed effect associated with the CC programme to vary across classes, assuming a random distribution of β_2 at class level. The model we are fitting is

$$\log\left(\frac{\gamma_{ijk}^{(s)}}{1 - \gamma_{ijk}^{(s)}}\right) = \alpha^{(s)} - \left(\sum_{l=1}^{4} \beta_l x_{lijk} + v_k + u_{0jk} + u_{1jk} x_{2ijk}\right), \qquad (8.11)$$

and the total variance at class level is modelled as a quadratic function of the effect of the CC programme with two variance terms and one covariance. However, fitting this model does not show any significant improvement over model C in Table 8.5. The variance associated with u_{1jk} is estimated to be zero, and the covariance between u_{0jk} and u_{1jk} is estimated to be -0.063, with standard error 0.059.

Table 8.6 Parameter estimates for the TVSFP data obtained by MIXOR and MLwiN.

Parameter	Model F	*Model G*	MIXOR
Fixed effect			
$\alpha^{(1)}$	−0.097 (0.148)	−0.085 (0.147)	0.076 (0.154)
$\alpha^{(2)}$	1.230 (0.150)	1.211 (0.149)	1.273 (0.063)
$\alpha^{(3)}$	2.398 (0.157)	2.439 (0.159)	2.479 (0.080)
β_1, Prethks	0.421 (0.039)	0.419 (0.039)	0.415 (0.041)
β_2, CC	0.863 (0.172)	0.870 (0.175)	0.861 (0.187)
β_3, TV	0.233 (0.170)	0.205 (0.171)	0.206 (0.168)
β_4, CC*TV	−0.342 (0.242)	−0.299 (0.246)	−0.301 (0.252)
Random effect			
σ_u^2		0.198 (0.059)	0.189 (0.076)
$\sigma_u^{2(1)}$	0.243 (0.093)		
$\sigma_u^{(1,2)}$	0.243 (0.071)		
$\sigma_u^{2(2)}$	0.219 (0.072)		
$\sigma_u^{(3,1)}$	0.182 (0.065)		
$\sigma_u^{(3,2)}$	0.148 (0.060)		
$\sigma_u^{2(3)}$	0.142 (0.070)		

Furthermore, we can fit a full variance–covariance structure at the class level associated with the estimates of the $\alpha^{(s)}$, assuming a different variance for each of them. The results are presented as model F in Table 8.6. It seems that the random effect estimate associated with $\alpha^{(1)}$ is the largest, whilst that of $\alpha^{(3)}$ is the smallest. However, the difference between the two is not statistically significant ($\chi_1^2 = 0.98$). We can rely on the simpler model with only one random effect estimate for all the ordinal series across classes.

8.3.5 MLwiN and MIXOR

If we fit model G in Table 8.6 and compare the estimates with those estimated by MIXOR, we find that the two programs give very similar estimates for the four covariates and their standard errors. The estimated random effect at class level also shows close agreement. However, the standard errors for $\alpha^{(2)}$ and $\alpha^{(3)}$ are considerably larger when estimated by MLwiN than those obtained by MIXOR. This may be due to the different ways in which the two programs treat the variance of the responses. In MLwiN the variance of the response vector for each student takes the form of (8.8), whilst in MIXOR it is taken to be equal to $\pi^2/3$.

8.3.6 A proportional hazards model

An alternative when modelling ordinal data is to use a proportional hazards model with a log–log link function, written as

$$\log\left[-\log\left(1 - \gamma^{(s)}\right)\right] = \alpha^{(s)} - \left(\sum_{l=1}^{4} \beta_l x_{lijk} + v_k + u_{jk}\right), \tag{8.12}$$

where the properties of the ordinal series of $\alpha^{(s)}$ remain the same as in model (8.9), but their interpretation and the estimates for the βs and the residuals are all different. If the proportion $\gamma^{(s)}$ is a measure related to a survival time, $1 - \gamma^{(s)}$ can then be considered the hazard. The double log transformation means that the $\alpha^{(s)}$ increase with a positive effect of the covariate. Hence an increase in $\alpha^{(s)}$ due to an effect of the covariate implies that this variable results in an increase in the hazard of being in category $s - 1$. For the dichotomous variable CC in x_2, its fixed effect estimate is β_2. The anti-log term e^{β_2} is interpreted as the ratio of the log hazard of being in the CC programme to the log hazard of not being in the programme (or in the base group), i.e. expressed as

$$e^{\beta_2} = \frac{\log\left(1 - \lambda^{(s)}\right)_{x_2=1}}{\log\left(1 - \lambda^{(s)}\right)_{x_2=0}}.$$

We can rewrite this expression as

$$\left(1 - \lambda^{(s)}\right)_{x_2=1} = \left(1 - \lambda^{(s)}\right)_{x_2=0}^{e^{\beta_2}}.$$

Fitting (8.12) to the data leads to the same conclusions on all parameter estimates in both the fixed part and the random part of the model. In Table 8.7, we compare the z-value (the estimate divided by its SE) between model G in Table 8.6 (the proportional odds model) and model (8.12) (the proportional hazards model).

Both of the linear transformations $\log[(\lambda^{(s)})/(1 - \lambda^{(s)})]$ and $\log[-\log (1 - \lambda^{(s)})]$ produce dependent variables ranging from $-\infty$ to $+\infty$. In most cases, the two models will give similar estimates on the same dataset.

Table 8.7 z-values from fitting the proportional odds and proportional hazards models.

Parameter	Model (8.12) (Log–log link)	Model G (Logit link)
Fixed effect		
β_1, Prethks	10.39	10.74
β_2, CC	5.06	4.97
β_3, TV	1.64	1.20
β_4, CC*TV	−1.66	−1.22
Random effect		
σ_u^2	2.82	3.36

ACKNOWLEDGEMENTS

The author wishes to thank Dr X.S. Li, Dr B.R. Flay, Dr B.R. Brannon and Dr D. Hedeker, who provided the data used in this chapter. This work has been performed with continuing support from the UK continuous Economic and Social Research Council (ESRC) to the Multilevel Models Project based at the Institute of Education, University of London.

CHAPTER 9

Institutional Performance

E. Clare Marshall and David J. Spiegelhalter

MRC Biostatistics Unit, Institute of Public Health, Cambridge UK

9.1 INTRODUCTION

Performance measures for institutions are becoming more and more central to efforts to introduce professional accountability and contain costs within health services (NHS Executive, 1995; Scottish Office, 1995; New York State Department of Health, 1995). Typically, the emphasis of such analyses lies not with the population as a whole: for example when analysing data on survival rates following kidney transplantation, one is not interested in the overall rate within the UK, say, but rather with the rates at each individual centre and the variability in rates between centres.

Health outcomes at different institutions will vary for at least three reasons: first, differences may be attributable to random variation; secondly, institutions may differ systematically with respect to the care they provide; and thirdly, the health of their respective patient populations may differ prior to admission. It is likely that variability in crude mortality rates, for example, will depend more on the mix of patients reaching hospital rather than on the care they receive once admitted. There is little doubt that the comparison of crude outcome rates is misleading (McKee and Hunter, 1995; Rowan *et al.*, 1993) and therefore that one should adjust for case-mix. 'Case-mix' refers to the factors that characterise the patient population in terms of diagnosis, age, sex, severity of disease etc., and 'case-mix adjustment' seeks to remove the impact of these factors, with the hope that any remaining differences in outcome between institutions, taking into account the play of chance, reflect 'quality of care'. Recent publications in which UK institutions have been explicitly compared in this way include those by de Courcy-Wheeler *et al.* (1995), who use the CRIB scoring system in a comparison of neonatal survival, and Rowan *et al.* (1993) who use the APACHE II (Knaus *et al.*, 1985) scoring system to adjust for case-mix when comparing survival rates in intensive care units.

Medical criticism of such profiling schemes has tended to concentrate on the inappropriate use of outcomes, such as mortality rates, as a sole measure of 'quality of care', the inadequacy of risk-adjustment procedures for varying case mix, and doubts about the reliability of the data *per se* (DuBois *et al.*, 1987; Jencks *et al.*, 1988; Epstein, 1995; Schneider and Epstein, 1996). We do not intend to add to this debate. Further controversy, however, surrounds the explicit ranking of institutions. This is generally avoided by those responsible for the profiling scheme, but consumer groups and the media almost inevitably publish 'league tables' of performance. As with any statistical summary measure, an institution's rank has associated uncertainty – uncertainty that must be taken into account before inferences regarding relative performance can be made.

This chapter has two broad objectives. The first is to show how multilevel modelling techniques can be applied to the analysis of clinical outcome data to produce more reliable estimates of performance. Specifically, these models overcome small sample problems by appropriately pooling information across institutions, introducing some bias or *shrinkage*, and providing a statistical framework that allows one to quantify and explain variability in outcomes through the investigation of institutional level covariates. Such models have been widely used in educational research (Bryk and Raudenbush, 1992; Aitkin and Longford, 1986; Goldstein *et al.*, 1993) but only recently have applications in medical comparisons been promoted (Thomas *et al.*, 1994; Normand *et al.*, 1995; Morris and Christiansen, 1996; Goldstein and Spiegelhalter, 1996; Rice and Leyland, 1996). The second is to investigate the value of an institution's rank as a reflection of relative performance. Recent computational advances, specifically in Markov-chain Monte Carlo methods (see the appendix to this chapter) allow one to quantify the uncertainty associated with an institution's rank and so to determine the extent to which conclusions may be based on explicit rankings.

The remainder of the chapter is organised as follows. The next section introduces the running example used throughout the chapter, comprising data annually published as part of an initiative by the New York State Department of Health to evaluate the performance of its hospitals and surgeons with respect to coronary artery bypass graft surgery. Section 9.3 considers the traditional fixed effects analysis where the surgeons are treated as independent; then Section 9.4 contrasts the results obtained with those assuming a two-level multilevel model. Section 9.5 looks at the surgeons' ranks under both models, with particular emphasis being paid to the uncertainty surrounding these ranks, while Section 9.6 extends the two-level model to include a third, hospital, level. Section 9.7 summaries our findings and discusses their implications for routine performance assessment. All analyses are carried out using the BUGS program (Spiegelhalter *et al.*, 1995), and computational issues are discussed in the appendix to the chapter.

9.2 NEW YORK DATA

The New York State Department of Health programme on cardiac artery bypass graft (CABG) surgery seeks to create a profiling system that monitors

Table 9.1 Observed, expected and risk-adjusted surgeon mortality after CABG surgery, 1991–1993 (from New York State Department of Health Publication, (1995).

	Cases	Number of deaths	Observed mortality rate, OMR	Expected mortality rate, EMR	Risk-adjusted mortality rate, RAMR	95% confidence interval for RAMR
Albany Medical Centre Hospital						
Bennett, E.	842	7	1.45	2.57	1.61	(0.65, 3.33)
Britton, L.	441	8	1.81	2.41	2.14	(0.92, 4.22)
Canavan, T.	510	14	2.75	2.32	3.38	(1.84, 5.66)
Ferraris, V.	341	11	3.23	3.46	2.66	(1.33, 4.76)
Foster, E.	227	8	3.52	3.16	3.18	(1.37, 6.26)
Arnot–Ogden Medical Centre						
Borja, A.	478	18	3.77	2.55	4.21	(2.49, 6.65)
Saifi	226	5	2.21	2.29	2.75	(0.89, 6.42)
Vaughan, J.	387	7	1.81	2.38	2.17	(0.87, 4.46)

the performance of hospitals and surgeons over time (New York State Department of Health, 1995, 1996). One of its explicit aims is to provide 'information to help patients make better decisions about referrals and treatment'. Table 9.1 shows a sample of the data published in 1995 covering operations for 1991–1993. Part of the published analysis comprises a logistic regression on the pooled data, adjusting for known risk factors of cardiac mortality. The resulting fitted probabilities, when added over a surgeon's cases, give an expected mortality adjusted for the severity of illness of his or her patients. The ratio of observed to expected mortality can be interpreted as the surgeon's standardised mortality rate, which, when multiplied by the state-wide average of 2.85%, provides a 'risk-adjusted mortality rate'. This latter quantity forms the basis for comparisons between individuals, and can be interpreted as the estimated mortality rate for a given surgeon had he or she operated on a population of patients identical to the statewide case-mix.

Let O_{ij} be the observed number of deaths in patients treated by surgeon i in hospital j and let E_{ij} be the number expected given the severity of their patients. The frequency of deaths during the three-year period can be assumed to have a Poisson distribution with unknown mean. That is,

$$O_{ij} \sim \text{Pois}(\mu_{ij}), \tag{9.1}$$

where

$$\log \mu_{ij} = \log E_{ij} + \theta_{ij}. \tag{9.2}$$

Unfortunately, we are not privy to the patient-level data, and so we use E_{ij} as a surrogate measure of severity. The idea is that E_{ij} adjusts for the patient effects, leaving θ_{ij} to represent the surgeon-specific effects of interest. It is convenient to think of the latter as log relative risks, although, strictly speaking, they are log standardised mortality rates. Following the published analysis, comparison between surgeons is based on a surgeon-specific risk-adjusted mortality rate, $2.85\,\exp(\theta_{ij})$.

9.3 A FIXED-EFFECTS MODEL – ASSUMING INDEPENDENT SURGEONS

We are adopting a Bayesian approach, and so need to select a 'minimally informative' independent prior for each θ_{ij}, which we choose to be

$$\theta_{ij} \sim N(0, 1 \times 10^6). \qquad (9.3)$$

This proper normal prior has a standard deviation of 1000, and since the observed log relative risks (even before any adjustment for patient severity) only range from 0 to 3.8, it seems reasonable that the prior only provides minimal information relative to that in the data. Adopting a normal prior for the log relative risks has the advantage that the model falls naturally into a hierarchical structure, allowing one to easily impose more complex forms of prior dependence (see later sections).

This fixed-effects analysis mirrors that published, although the intervals surrounding the estimates of risk adjusted mortality rates do differ slightly. Here we use Bayesian posterior credible intervals, whereas New York State Department of Health (1995) quotes those obtained using the approximation of Rothman and Boice (1979). The latter tend to be slightly conservative.

Figure 9.1 shows the estimated risk adjusted mortality rates, along with associated intervals under this above fixed effects model. The estimated risk-adjusted mortality rate for each surgeon, together with the number of operations they performed and a hospital indicator, are shown in parentheses. Figure 9.1 suggests substantial variability in outcome between surgeons, with observed risk-adjusted mortality rates ranging from 0.22% to 8.02%. In the New York programme, a surgeon is considered to be aberrant and is isolated for further review if the interval associated with the estimate of his or her risk-adjusted mortality rate does not include the statewide average of 2.85%. Following this rationale 16 surgeons would be picked out: 5 for having 'significantly' high adjusted mortality rates and 11 for having 'significantly' low adjusted mortality rates. Note, however, that four surgeons would be expected to fulfil this criterion by chance alone.

Another feature to notice is that those surgeons performing a large number of CABGs over the period of study appear to have lower adjusted mortality rates.

9.4 A TWO-LEVEL MODEL

We now consider an alternative approach that in some sense pools information across surgeons to obtain more reliable estimates of performance. Here we consider a two-level model. By explicitly modelling the heterogeneity between

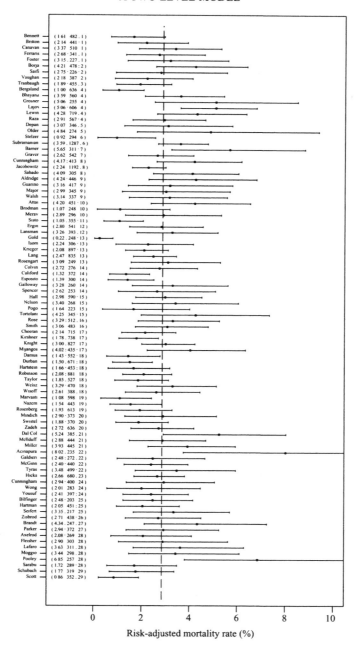

Figure 9.1 Mean and 95% intervals for the risk-adjusted mortality rate for each surgeon over the period 1991–1993, assuming the independence of surgeons (the state-wide average of 2.85% is shown). The estimated risk-adjusted mortality rate of each surgeon, together with the number of operations they performed and a hospital indicator, are given in parentheses

surgeons, the aim is to shed light on the underlying distribution of mortality rates, and so help identify surgeons with 'genuine' high or low mortality rates, compensating for the fact that 'smaller' surgeons are more likely to be outlying by chance alone. The joint modelling should also help to overcome the classical problem of multiple comparisons – encountered here since we are comparing many outcomes with a standard.

The surgeon effects are assumed to be exchangeable (Bernardo and Smith, 1994), and we begin by assuming that the distribution of surgeons has the simplest of functional forms, drawn from a common normal distribution. By assuming exchangeability, we are acknowledging the heterogeneity between surgeons but are assuming that this heterogeneity is unstructured, i.e. it is not a priori predictable (this assumption is relaxed later). Thus

$$\theta_{ij} \sim N(\alpha, \sigma^2). \tag{9.4}$$

Here $2.85\exp(\alpha)$ represents the underlying overall adjusted mortality rate and σ^2 gives a measure of the between-surgeon variability (on a log scale). Vague but proper priors are specified for the hyperparameters α and σ^2:

$$\alpha \sim N(0, 1 \times 10^6), \tag{9.5}$$

$$\sigma^{-2} \sim \Gamma(0.001, 0.001). \tag{9.6}$$

The gamma distribution for the precision (inverse of the variance) has a mean of 1 and variance of 1000: this 'just proper' prior essentially puts a locally uniform distribution on the logarithm of the variance.

Figure 9.2 shows the estimates and associated intervals for each surgeon's adjusted mortality rate under this two-level model. Comparing this with Figure 9.1 clearly shows the shrinkage toward a global mean, with those surgeons performing fewer operations over the period of study being shrunk more than those performing a large number. This seems reasonable since estimates based on large populations are preserved yet those based on more unreliable data are shrunk – the shrunk estimates have the appealing property of falling between those obtained under the two extremes of independence and the complete pooling of surgeons. Nine of the 16 surgeons who would have been isolated for further review under the fixed-effects model are no longer considered to be aberrant. For example, consider Pooley (hospital 28), who, under the approach adopted by the New York State Department of Health would have been considered to have an unacceptable high observed mortality rate, no longer fulfils that criterion. Gold (hospital 13), on the other hand, has a risk adjusted mortality rate that is 'significantly' below the standard even after assuming the two-level model.

9.5   RANKS FOR INSTITUTIONS

No matter how careful the statistical analysis, it is almost inevitable that the resulting estimates will lead to institutional ranking and 'league tables'. For

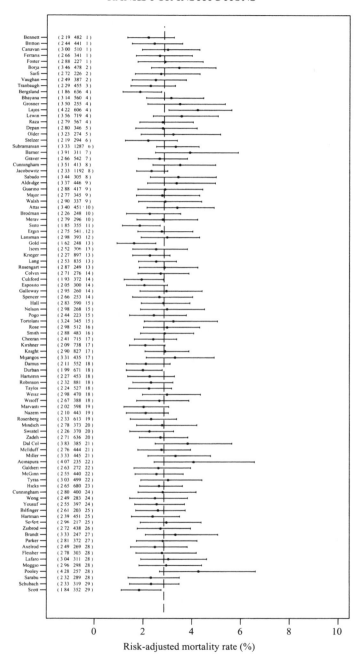

Figure 9.2 Mean and 95% intervals for the risk-adjusted mortality rate for each surgeon, assuming exchangeable surgeons (the state-wide average of 2.85% is shown). The estimated risk-adjusted mortality rate for each surgeon, together with the number of operations they performed and a hospital indicator, are given in parentheses

example, as part of the Health of the Nation programme (NHS Executive, 1995) every Health Authority in England and Wales was assessed to establish its relative performance according to certain indicators, such as the annual rates of strokes or coronary heart disease. The graphics in publications associated with this programme consistently showed regions in rank order for all outcomes with no measure of uncertainty.

Although such ranks are extremely sensitive to sampling variation it is not easy within the classical framework to quantify their uncertainty. This is straightforward, however, if one adopts a Markov-chain Monte Carlo (MCMC) approach (see the appendix at the end of this chapter for an introduction to the ideas of MCMC).

At each iteration of the Gibbs sampler, used in our analysis, not only are samples generated from the posterior distribution of each surgeon's adjusted mortality rate, but this set of parameter realisations are ranked. Suppose, for example, that the Gibbs sampler were run for N iterations; then, discarding an initial burn-in of n, we are left with a sample of size $N - n$ of each surgeon's adjusted mortality rate and a sample of size $N - n$ of each surgeon's rank. Point and interval estimates of the ranks can then be reported alongside the estimates of performance (Marshall and Spiegelhalter, 1998).

Figures 9.3 and 9.4 show the 95% intervals associated with each surgeon's rank under the fixed and multilevel models (the lower the rank the better). The intervals around the ranks are wide – particularly those assuming a multilevel model, where only two surgeons – Lajos (hospital 4) and Pooley (hospital 28) – could be confidently placed in the bottom half, and only Gold is consistently ranked in the top half.

9.6 A THREE-LEVEL MODEL

The organisational structure or practice style of a hospital may result in an association of outcomes within hospitals even after accounting for patient and surgeon effects. We incorporate this into the model by considering an elaboration of the two-level model to include a third, hospital, level. The model assumes the exchangeability of surgeons within the same hospital but the conditional independence of surgeons at different hospitals. We have

$$\theta_{ij} \sim N(\lambda_j, \tau^2), \tag{9.7}$$

$$\lambda_j \sim N(\eta, \omega^2). \tag{9.8}$$

The λ_j, $j = 1, \ldots, 29$, represent the underlying log hospital-specific relative risks. Common within-hospital variability τ^2 is assumed. The surgeon effects of interest are now given by the level-2 residuals $\rho_{ij} = \theta_{ij} - \lambda_j$, and surgeons are compared via $2.85\exp(\hat{\rho}_{ij})$, which is the underlying estimated mortality rate for surgeon i, hospital j, adjusting for both the severity of the patients and the hospital he or she is working in. Vague but proper priors are specified for the hyperparameters (τ^2, ω^2, η).

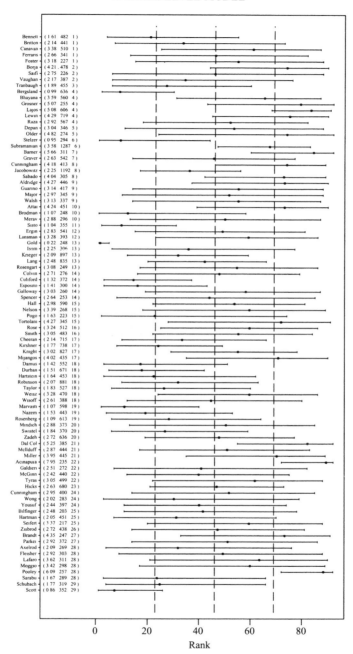

Figure 9.3 Mean and 95% intervals for the rank of each surgeon, assuming independent surgeons. The estimated risk-adjusted mortality rate for each surgeon, together with the number of operations they performed and a hospital indicator, are given in parentheses

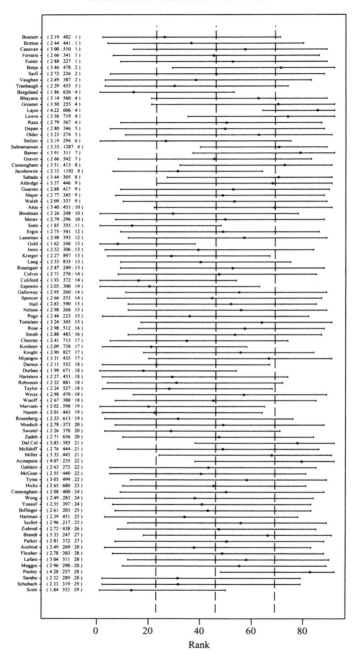

Figure 9.4 Mean and 95% intervals for the rank of each surgeon, assuming exchangeable surgeons. The estimated risk-adjusted mortality rate for each surgeon, together with the number of operations they performed and a hospital indicator, are given in parentheses.

Table 9.2 Estimated variance components under two- and three-level models.

Model	Estimate (SE)
Two-level model:	
σ^2 (between-surgeon)	0.081 (0.029)
Three-level model:	
τ^2 (between-surgeon–within-hospital)	0.063 (0.028)
ω^2 (between-hospital)	0.021 (0.019)

Table 9.2 gives the estimate of the between-surgeon variability obtained through the fitting of the two-level model, along with the between-surgeon–within-hospital and between-hospital variabilities assuming the three-level model described here. Note that $\hat{\sigma}^2 \approx \hat{\tau}^2 + \hat{\omega}^2$. The results show that only a quarter of the between-surgeon variability estimated under the two-level model can actually be attributed to variability between hospitals. Once this is modelled explicitly, no surgeons have estimated risk-adjusted mortality rates that are significantly different from the state-wide average (see Figure 9.5). The plot of ranks (Figure 9.6) also suggests substantial homogeneity, in spite of the 'significant' variance τ^2, with only one surgeon being ranked consistently in the latter half.

It is interesting to focus once again on Pooley and Gold. Pooley's estimated risk-adjusted mortality rate is not affected by the inclusion of a hospital level – suggesting that the hospital he is working at (hospital 29) is not atypical. This is the case, since hospital 29 has an estimated centre-specific risk-adjusted mortality rate of 2.85 (0.33) – exactly the statewide average. The interval surrounding Pooley's rank, however, is affected by the elaboration of the model and is now substantially wider. After adjusting for a hospital effect, Gold's estimated risk-adjusted mortality rate is no longer significantly below the statewide average. This suggests that he is a surgeon working at a 'good' hospital.

9.7 DISCUSSION

One crucial issue, not explored here, is that concerning the sensitivity of inferences to different model assumptions. All models considered in this chapter have relied on the explicit assumption that the institutional (surgeon or hospital) effects are drawn from a normal distribution. This needs justification. Research into methods of criticising multilevel models is very much ongoing. For example, issues of outlier detection are addressed by Langford and Lewis (1998), while Lange and Ryan (1989) investigate the use of normal quantile–quantile plots to assess the appropriateness of the normality assumption for second-level parameters. Issues of how to incorporate uncertainty effectively in these plots, however, remain unresolved.

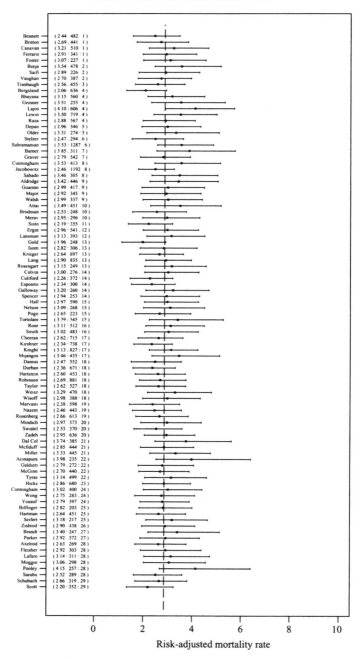

Figure 9.5   Mean and 95% intervals for the risk-adjusted mortality rates – assuming a three-level model. The estimated risk-adjusted mortality rate for each surgeon, together with the number of operations they performed and a hospital indicator, are given in parentheses.

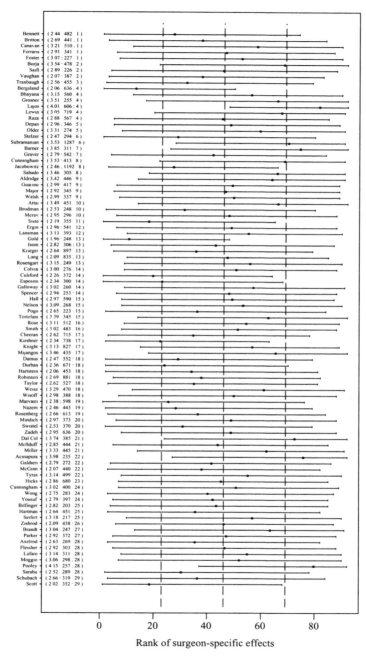

Figure 9.6 Mean and 95% intervals for the rank of each surgeon – assuming a three-level model. The estimated risk-adjusted mortality rate for each surgeon, together with the number of operations they performed and a hospital indicator, are given in parentheses.

If the normal distribution is not appropriate, a Markov-chain Monte Carlo approach allows full flexibility in the choice of the distributional form. For example, heavy-tailed distributions, or outlier or mixture models, can easily be fit at any level of the hierarchy (for example, see Albert and Chib, 1997; Richardson and Green, 1997).

It is clear that any attempt at using ranks to compare surgeons may be seriously misleading. This is only to be expected, since the bulk of surgeons have overlapping intervals, and hence precision in ranking is rarely obtainable, particularly for 'smaller' surgeons. The strength of recent statistical developments is to quantify this lack of precision, and hence emphasise the caution with which any 'league tables' must be treated.

APPENDIX: MCMC AND BUGS

Markov-chain Monte Carlo (MCMC) describes the approach of using a sample comprising of realisations of a Markov chain to obtain a Monte Carlo estimate of an integral. Essentially, Monte Carlo integration approximates $E_{\theta|Y}[f(\theta)] = \int f(\theta)p(\theta|Y)d\theta$, say, by the mean of $\{f(\theta^{(k)})\}$, where $\{\theta^{(k)}; k = 1, \ldots, K\}$ represents a sample from $p(\theta|Y)$. One way of drawing this sample is through a Markov chain that has $p(\theta|Y)$ as its stationary distribution. Constructing such a chain is often surprisingly easy. The BUGS software uses one particular method known as the Gibbs sampler (Geman and Geman, 1984); for a more detailed introduction to MCMC, see Gilks *et al.* (1996) and Gelfand and Smith (1990).

The BUGS software (Spiegelhalter *et al.*, 1995) automatically generates the code to carry out the Gibbs sampling by exploiting the conditional independence assumptions implicit in the model specification. Consider, for example, the three-level model described earlier, a graphical representation of which is given in Figure 9.7. Each node of this graph represents a quantity in the model.

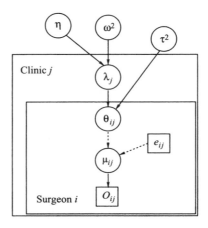

Figure 9.7 Graphical representation of the three-level model.

Gibbs sampling works by iteratively sampling from the full conditional distributions of the unobserved nodes in the graph. The full conditional distribution of a node is defined as the distribution of that node given the current or known values of all other nodes in the model. Exploiting the conditional independence structure of the graph simplifies these full conditionals, making the sampling relatively straightforward (see Spiegelhalter (1998) for an explanation of graphical modelling, and Ripley (1987) for a description of sampling methods). The BUGS code for this model is shown below. There is not a direct translation between the mathematical or graphical specification and the BUGS code in this example, but nevertheless the syntax should be fairly self-explanatory. When the data are unbalanced – as here, where we have variable numbers of surgeons at each hospital – it is easier to assume a single index for the observations and code hospital as a factor. Another point to note is that BUGS describes normal distributions in terms of their mean and their precision.

The BUGS software can be freely downloaded from
http://www.mrc-bsn.cam.ac.uk/bugs

The BUGS code

```
model 3level;
const
   N=92,               #number of surgeons
   K=29;               #number of hospitals
var
   o[N],               #Observed deaths for ith surgeon.
   e[N],               #Expected deaths for ith surgeon given case mix.
   mu[N],
   theta[N],
   hospital[N],        #Hospital indicator for ith surgeon
   delta,              #Between-surgeon-within-hospital precision
   surg[N],            #Surgeon-specific effect
   lambda[K],          #Log hospital specific relative risk
   r[N],               #Surgeon-specific risk adjusted mortality rate
   eta,                #Estimated overall log relative risk
   tau.sq,             #Between-surgeon-within-hospital variance
   gamma,              #Between hospital precision
   omega.sq,           #Between-hospital variance
   not.greater[N,N],
   surgeon.rank[N];

data in "data.dat";
inits in "initials.in";

{
for(i in 1:N){
```

```
  o[i] ~ dpois(mu[i]);
  log(mu[i]) <-log(e[i])+theta[i];
  theta[i] ~ dnorm(lambda[hospital[i]],delta);
  surg[i] <-theta[i]-lambda[hospital[i]];
  r[i] <-2.85*exp(surg[i])

#Compute ranks:
  for(j in 1:N){
     not.greater[i,j] <-step(r[i]-r[j]);
  }
  surgeon.rank[i] <-sum(not.greater[i,]);
  }

for(j in 1:K){
  lambda[j]~dnorm(eta,gamma);
  }
#Priors:
delta ~ dgamma(1.0E-3,1.0E-3);
tau.sq <-1/delta;
eta ~ dnorm(0.0,1.0E-6);
gamma ~ dgamma(1.0E-3,1.0E-3);
omega.sq <-1/gamma
}
```

CHAPTER 10

Spatial Analysis

Alastair H. Leyland

MRC Social and Public Health Sciences Unit, University of Glasgow, UK

10.1 INTRODUCTION

In a recent review of the rationale for the inclusion of geography in a statistical analysis, Wakefield and Elliott (1999) drew a distinction between four types of study: disease mapping, geographic correlation studies, the assessment of risk in relation to a point or line source, and the issues surrounding cluster detection and disease clustering. Disease mapping is an established tool in public health, and has the main purpose of the removal of random noise and artefacts of population variation (Lawson *et al.*, 1999b) – a form of image processing leading to a map that better reflects, for example, the geographical distribution of the relative risk of a disease. Geographic correlation studies, also known as ecological studies, seek to describe the relationship between the geographical variation in a disease and the variation in exposure to a factor, possibly a hypothesised cause of disease (English, 1992). Where exposure is believed to emanate from a known point or line source – often through some form of pollution – then the risk should increase with proximity to the source, and statistical models should take the location of the source into account (Lawson *et al.*, 1999a). Cluster detection can be used as a form of surveillance, with the intention being early detection of heightened risk of disease in certain areas with no prior hypothesis (Elliott *et al.*, 1995). This chapter concentrates on the first two of these, namely disease mapping and geographic correlation studies, and considers the analysis of counts of disease within small areas rather than the analysis of the location of each case.

The following section describes a spatial model in a multilevel structure, assuming that the rates of disease incidence or mortality are sufficiently rare for a Poisson model to be appropriate, and concentrates on an autoregressive error structure. Section 10.3 considers the computational aspects of such models, whilst Section 10.4 analyses data on lip cancer incidence in Scotland. Section 10.5 describes how the general multilevel structure can readily encompass higher-order autoregressive terms, additional geographical levels,

Multilevel Modelling of Health Statistics Edited by A.H. Leyland and H. Goldstein
©2001 John Wiley & Sons, Ltd

multivariate spatial models and spatio-temporal models. Section 10.6 contains some discussion of these models and the conclusions of the chapter.

10.2 THE SPATIAL MODEL

A number of models have been proposed for the analysis of geographically structured data reflecting disease risk (Bernardinelli and Montomoli, 1992). The family of models considered in this chapter assume that the observed number of events – such as deaths – in area i, O_i, follow a Poisson distribution (see Chapter 4) with mean μ_i, so that

$$O_i \sim \text{Poisson}(\mu_i). \tag{10.1}$$

For each area E_i, the number of deaths that would be expected were national age- and sex-specific death rates to apply to the population of area i can be estimated. The mean number of deaths in each area, μ_i, can be written in terms of this expected number of deaths and the (unknown) underlying relative risk for the ith area, ξ_i:

$$\mu_i = E_i \xi_i. \tag{10.2}$$

The maximum-likelihood estimate (MLE) of the relative risk is

$$\hat{\xi}_i = O_i/E_i, \tag{10.3}$$

which is equivalent to the standardised mortality ratio (SMR). The SMR does, however, have a number of drawbacks; in particular, it is imprecisely estimated when the number of cases is small (Clayton and Kaldor, 1987). This has the consequence that maps showing SMRs tend to be dominated by areas with small populations – which are often large geographical areas – with SMRs that are estimated to be extremely high or extremely low.

An alternative is to model the logarithm of the relative risk in terms of known covariates x_i and unknown area effects ϕ_i, so that the logarithm of the mean number of deaths in area i can be written as

$$\log \mu_i = \log E_i + x_{0i}\beta_0 + \ldots + x_{pi}\beta_p + \phi_i. \tag{10.4}$$

The logarithm of the expected number of deaths is assumed to be known, and is treated as an offset in this model (see Chapter 4), and the β are fixed effects to be estimated as usual. A simple model of area heterogeneity assumes independence of the area effects; this can be written as

$$\phi_i \sim N(0, \sigma^2). \tag{10.5}$$

The area effects ϕ_i produced under such a model are shrunk towards a *global* mean effect of zero; the relative risks, however, will not be shrunk towards an overall mean in the presence of covariates. This ignores the geographical structure of the data; it may be, for example, that there is noticeable clustering of areas with high risk and areas with low risk above that which may be predicted by the covariates. An alternative is therefore to consider some form of *local clustering* or *shrinkage*; a commonly used example is a conditional autoregressive model under which, conditioning on the other area effects $\{\phi_j\}$, the distribution of the effect for the ith area is given by

$$\phi_i \big| \{\phi_j\} \sim N\left(\bar{\phi}_i, \sigma_{\phi i}^2\right), \tag{10.6}$$

where $\bar{\phi}_i$ denotes the mean of the area effects in those areas neighbouring area i. (The mean of the neighbouring areas is not the only summary that could be used; in principle, it would be possible to choose other parameters of the distribution of the $\{\phi_j\}$.) In general and for irregular maps – those in which the number of neighbours n_i is not constant for each area – the variance term will decrease as n_i increases; a common choice is $\sigma_{\phi i}^2 = \sigma_\phi^2 / n_i$.

It is frequently desirable to fit a model that combines an exchangeable global clustering element with structured local clustering. One possibility is to consider a weighted sum of (10.5) and (10.6); Langford *et al.* (1999a, b) developed a generalised multilevel formulation following an approach suggested by Besag *et al.* (1991) under which the area effect ϕ_i is written as the sum of a heterogeneity component μ_i and an independent spatially structured component v_i:

$$
\begin{aligned}
\phi_i &= u_i + v_i, \\
u_i &\sim N\left(0, \sigma_u^2\right), \\
v_i \big| \{v_j\} &\sim N\left(\bar{v}_i, \sigma_{vi}^2\right).
\end{aligned}
\tag{10.7}
$$

The alternative structure suggested by Langford *et al.* (1999a, b) was to write

$$
\left.
\begin{aligned}
\phi_i &= u_i + v_i, \\
v_i &= \sum_{j \neq i} z_{ij} v_j^*, \\
u_i &\sim N\left(0, \sigma_u^2\right), \\
v_j^* &\sim N\left(0, \sigma_v^2\right), \\
\mathrm{cov}\left(u_i, v_i^*\right) &= \sigma_{uv},
\end{aligned}
\right\}
\tag{10.8}
$$

so that, unlike model (10.7), the spatial and heterogeneity effects are allowed to be correlated. Here the z_{ij} are spatial explanatory variables, and represent a measure of the relevance of area j to area i. Choices of z_{ij} considered include models of exponential decay based on the distance between area centroids – effectively smoothing across the entire map, but more towards nearby areas than distant areas – and an autoregressive model such that $z_{ij} = 1/n_i$ if areas i and j are neighbours, 0 otherwise. All of the models considered in the rest of this chapter assume such an autoregressive structure. As such, the heterogeneity model described by (10.5) can be seen to be a special case when $\sigma_v^2 = \sigma_{uv} = 0$; moreover, a simple autoregressive model arises when $v_i^* = u_i$, i.e. when $\sigma_u^2 = \sigma_v^2 = \sigma_{uv}$. The final model for the unknown area effects ϕ_i is again such that the variance decreases as the number of neighbours n_i increases:

$$\phi_i \sim N\left(0, \sigma_u^2 + \sigma_v^2/n_i\right) \qquad (10.9)$$

The covariance between two areas i and j is $\left(1/n_i + 1/n_j\right)\sigma_{uv} + \left(n_{ij}/n_i n_j\right)\sigma_v^2$ if i and j border each other, where n_{ij} is the number of common neighbours. If i and j do not border each other then the first term disappears and the covariance simplifies to $\left(n_{ij}/n_i n_j\right)\sigma_v^2$; this means that there is no covariance between areas that have no common neighbours.

The variance of the ϕ_i has two components: one arises through heterogeneity and the other through spatial structure. An approximation of the relative importance of the spatial part of the model is therefore given by the ratio $\sigma_v^2/\bar{n}\sigma_u^2$, where $\bar{n}$ is a suitable average of the number of neighbours of each area such as the mode. A value of one for this ratio suggests that the two components are of equal importance. As suggested by Besag *et al.* (1991), it will frequently be the case that either the spatial structure or the heterogeneity effects will dominate but which one is unlikely to be known in advance.

10.3   COMPUTATION

This chapter considers fitting spatial models using quasi-likelihood. A fully Bayesian alternative is Markov-chain Monte Carlo (MCMC) estimation (Mollié, 1996) using appropriate software (Spiegelhalter *et al.*, 1995). Comparisons between methods have been published (Bernardinelli and Montomoli, 1992); Section 10.6 includes some discussion of this topic.

Estimation of the model detailed by (10.1), (10.4) and (10.8) under an iterative generalised least-squares (IGLS) framework has been detailed by Langford *et al.* (1999a), with particular emphasis on the penalised quasi-likelihood (PQL) estimation used to approximate the Poisson distribution. This enables the user to fit such models using multilevel modelling software such as MLwiN (Rasbash *et al.*, 1999a,b), and has the advantage of simplifying the model extensions detailed in Section 10.5. The heterogeneity models and simple autoregressive models can be fitted by imposing the constraints detailed in Section 10.2 upon this general model. It is also possible to fit a spatial model conditioning on the residuals (10.6); at the k_{th} iteration the spatially-determined function of the area residuals from the previous iteration can be included in the offset, effectively centring each area around the mean of its neighbours:

$$\left.\begin{aligned}
\log \mu_i^{(k)} &= \left(\log E_{ij} + \sum_{j\neq i} z_{ij} u_j^{(k-1)}\right) + x_{0i}\beta_1^{(k)} + \ldots + x_{pi}\beta_p^{(k)} + u_i^{(k)}, \\
u_i &\sim N(0, \sigma_i^2).
\end{aligned}\right\} \qquad (10.10)$$

By constantly updating the offset, this model can be fitted to convergence, i.e. until $u_i^{(k)} = u_i^{(k-1)}$. A model of $\sigma_i^2 = \sigma^2$ fits a constant variance to all areas; alternatively, $\sigma_i^2 = \sigma^2 \sum_i n_i/mn_i$, where m is the total number of areas under study, weights the variance according to the number of neighbours.

10.4   EXAMPLE: SCOTTISH LIP CANCER DATA

As an example, we can consider registrations of cases of lip cancer in Scotland over the period 1975–1980, a data set first analysed by Clayton and Kaldor (1987). These data are recorded for the 56 local government districts in Scotland, and the full data set, as published for example in Stern and Cressie (1999), includes not just observed and expected counts for each district but also a covariate indicating the percentage of the population employed in agriculture, fishing and forestry as a measure of exposure to sunlight and a potential risk factor.

Figure 10.1 shows the crude standardised morbidity ratios (SMRs) for each district. This map indicates that there is a tendency for areas to cluster, with a noticeable grouping of areas with particularly high SMRs (>200) to the North of the country and in the islands. Standardised morbidity ratios range from 0 (two areas, Annandale and Tweeddale, had no notifications of lip cancer during the six years) to 652. These estimates are, however, based on extremely small numbers; the three areas at the extremes have a total of 7.3 expected deaths between them, whereas there were on average 9.6 deaths *per area*. Figure 10.2 then displays the SMRs estimated under a global clustering model, with changes from Figure 10.1 reflecting a combination of the uncertainties associated with small numbers in some areas and the inclusion of additional information (the percentage of agricultural, fishing and forestry workers). The estimated SMRs now range from 33 to 446. The small population sizes in the areas with no observed events – which in turn give rise to low expected numbers of events – resulted in their estimates being shrunk to 61 and 71.

Table 10.1 displays parameter estimates for four models; the model of global clustering is labelled model A. The interpretation of the parameter estimates is as suggested for the Poisson models in Chapter 4. The percentage of agricultural, fishing and forestry workers has been centred around its (unweighted) mean of 8.39%; across Scotland, the range of this variable is from 0% to 24%. The parameter estimate in Table 10.1 suggests that this crude measure of exposure would on average lead to a relative risk in the areas with the highest proportion of such workers (Berwickshire, Orkney, Stewartry and Wigtown), which was 4.76 ($= \exp(24 \times 0.065)$) times greater than in areas with no agricultural, fishing and forestry workers (Bearsden, Glasgow, Inverclyde and Motherwell). The estimated variance can be used to provide coverage intervals for the relative risks, and we expect that 95% of areas have relative risks of between 0.29 and 3.44 ($= \exp(\pm1.96 \times \sqrt{0.397})$) above the risk associated with the proportion of outdoor workers. Figure 10.2 displays the estimated SMRs for each district under the global clustering model.

Model B in Table 10.1 is the model conditioning on the residuals given by equations (10.1), (10.4) and (10.6), under which the mean for each area is centred on the mean of its neighbours. This produces a reduction both in the effect of the percentage of workers in agriculture, fishing and forestry (with the estimated relative risk of lip cancer between the high- and low-exposure areas in the above example being just 2.87) and in the between area variance. Having

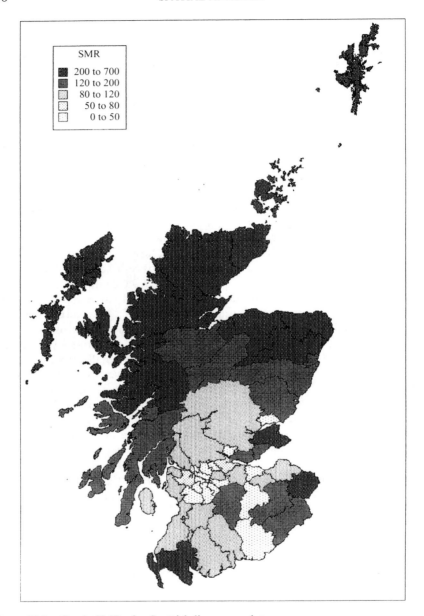

Figure 10.1   Crude SMRs for Scottish lip cancer data.

controlled for the effect of exposure, 95% of areas have relative risks of between 0.36 and 2.76 times the mean of the surrounding areas. Conditioning on the residuals of the surrounding areas has reduced the variance by about 40%; the estimated SMRs for each area are mapped in Figure 10.3. However, the

Table 10.1 Parameter estimates for global clustering (model A), model conditioning on the residuals (model B), simple autoregressive model (model C) and unconstrained spatial model (model D).

	Model A		Model B		Model C		Model D	
	Estimate	SE	Estimate	SE	Estimate	SE	Estimate	SE
Fixed								
Constant	0.082	0.101	0.023	0.085	0.079	0.151	0.107	0.161
% workers in agriculture, fishing and forestry	0.065	0.015	0.044	0.013	0.068	0.022	0.070	0.024
Random								
σ_u^2	0.397	0.105	0.236	0.070	0.352	0.138	0.165	0.363
σ_{uv}					0.352	0.138	0.296	0.306
σ_v^2					0.352	0.138	0.781	0.932

variance also decreases as the number of neighbours of an area (and hence the information available) increases; the value in Table 10.1 corresponds to an area with 4.71 neighbours (the mean) and varies in this data set from 0.101 for an area with 11 neighbours (Perth and Kinross) to 1.113 for an area with just one neighbour (Shetland).

Model C is the simple autoregressive model described by (10.1), (10.4) and (10.8) under the constraint that $\sigma_u^2 = \sigma_v^2 = \sigma_{uv}$. This again centres each area's estimated relative risk around the mean of the surrounding areas. Unlike model B, model C does not condition on the relative risks of neighbouring areas, but does take into account the uncertainty associated with their estimation. The estimated effect of the percentage of agricultural, fishing and forestry workers is approximately the same as in model A, and the estimated variance for an area with the mean number of neighbours – and which is therefore comparable with the variances for the other two models – is 0.427. This will vary from 0.384 for an area with 11 neighbours to 0.705 for an area with one neighbour. The different SMRs obtained under this model are mapped in Figure 10.4. Correlations between areas may be estimated as detailed in Section 10.2; the correlation between the relative risks for two areas each with five neighbours, two of which are common neighbours, will be 0.40 if the areas border each other and 0.07 if they have no common boundary. The constraint on the variances, ensuring that $v_i^* = u_i$, means that in terms of (10.9) the heterogeneity effects exert n_i times the influence of the spatial effects, or 4.71 times on average.

Model D is the full spatial model again given by (10.1), (10.4) and (10.8), but this time with no constraints on the parameters. The effect of the percentage of workers in agriculture, fishing and forestry is approximately the same as under models A and C. The estimated variance for an area with the mean number of neighbours is 0.331, and ranges from 0.236 for an area with 11 neighbours to

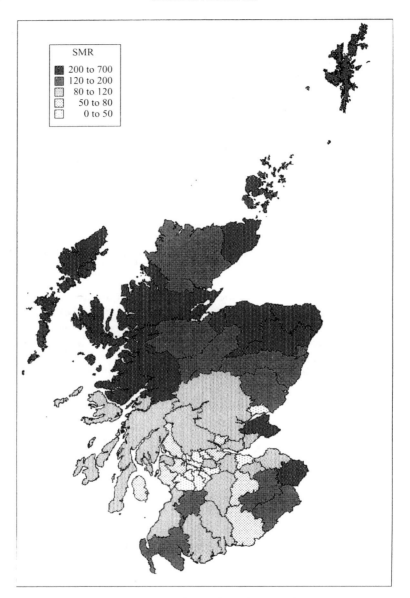

Figure 10.2 Lip cancer SMRs under global clustering model.

0.946 for an area with just one neighbour. The heterogeneity effects and the spatial effects are highly correlated – from Table 10.1, this correlation is estimated to be 0.82 – implying that the relative risk will tend to be higher in areas neighbouring an area with a high relative risk. The unconstrained estimation of the u_i and the v_i has resulted in an increased weight being placed on the spatial effects, which now exert almost exactly the same influence on the

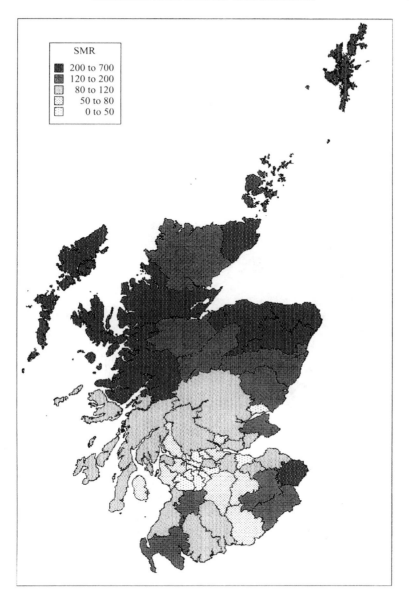

Figure 10.3 Lip cancer SMRs under model conditioning on residuals.

total variation as the heterogeneity effects. The equivalent correlation between the two areas used as an illustration in the previous paragraph is now estimated to be 0.56 if they border each other and 0.19 if they are not contiguous; the increases over the estimates obtained under model D are again testimony to the increased importance given to the spatial component. The fully smoothed map is given in Figure 10.5.

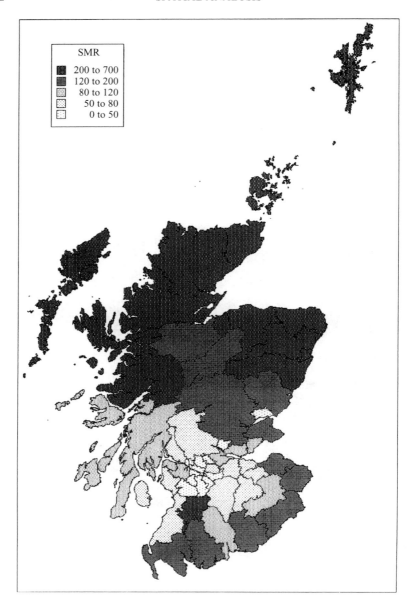

Figure 10.4   Lip cancer SMRs under simple autoregressive model.

10.5   EXTENSIONS TO THE SPATIAL MODEL

This section discusses how the spatial model described by (10.1), (10.4) and (10.8) can be extended in a fairly straightforward manner, leading to more realistic modelling of complex spatial processes.

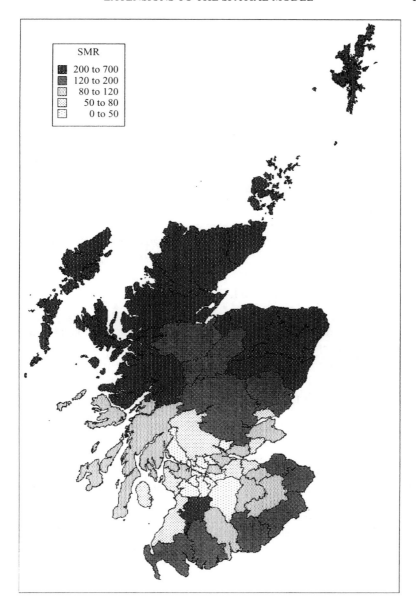

Figure 10.5 Lip cancer SMRs following full spatial smoothing.

10.5.1 Higher-order autoregressive models

Further autoregressive terms can be added by adding additional random effects in a similar manner to the spatial effects v_j^*. If (10.8) is rewritten such that

$$\phi_i = u_i + v_i + w_i \tag{10.11}$$

then the w_i can be defined in terms of random effects w_j^*, related through a relevance measure z_{wij}, such that

$$\left. \begin{aligned} w_i &= \sum_{j \neq i} z_{wij} w_j^*, \\ w_i^* &\sim N\!\left(0, \sigma_w^2\right), \\ \operatorname{cov}\!\left(u_i, w_i^*\right) &= \sigma_{uw}, \\ \operatorname{cov}\!\left(v_i^*, w_i^*\right) &= \sigma_{vw}. \end{aligned} \right\} \qquad (10.12)$$

The z_{wij} can incorporate any further measure of area relationships, for example indicating areas that do not border each other but have at least one common neighbour. The six random parameters (three variances and three covariances) may be estimated as before, and the variances of and covariances between areas can be expressed in terms of these parameters. Note that two areas will now be correlated if they share a common second neighbour, i.e. there may be up to three areas between them.

10.5.2 Models with three or more levels

Further levels can be added to spatial multilevel models as they can to any other multilevel model; moreover, the geographical dependence between units at any level of analysis can be modelled. So, for example, considering the effect of the jth country φ_j in addition to that of the ith region ϕ_{ij}, (10.4) becomes

$$\log \mu_{ij} = \log E_{ij} + x_{0ij}\beta_0 + \ldots + x_{pij}\beta_p + \phi_{ij} + \varphi_j. \qquad (10.13)$$

The country effects may be considered to be exchangeable with

$$\varphi_j \sim N\!\left(0, \sigma_\varphi^2\right). \qquad (10.14)$$

Such a model is appropriate if, for example, cultural or behavioural differences or differences in reporting events between countries are likely to lead to differences in the outcome, but if there is no geographical pattern to such behaviour. However, if the differences between countries are not random but exhibit a spatial pattern over and above that at the regional level then the country effects could be modelled as

$$\left. \begin{aligned} \varphi_j &= u_{\varphi,j} + v_{\varphi,j}, \\ u_{\varphi,j} &\sim N(0, \sigma_{\varphi,u}^2), \\ v_{\varphi,j} &= \sum_{k \neq j} z_{\varphi,jk} v_{\varphi,k}^*, \\ v_{\varphi,k}^* &\sim N\!\left(0, \sigma_{\varphi,v}^2\right), \\ \operatorname{cov}\!\left(u_{\varphi,j}, v_{\varphi,j}^*\right) &= \sigma_{\varphi,uv}. \end{aligned} \right\} \qquad (10.15)$$

10.5.3   Multivariate spatial models

Just as the geographical location of a particular area may be used to inform the estimated outcome for that area – based on empirical evidence or prior belief – so information may be obtained from other sources. One such possibility is where more than one response is available for each area, and where these responses are correlated. (See Chapter 5 for a discussion of multilevel multivariate regression models.) Leyland *et al.* (2000) give an example of the use of a multivariate spatial model for two broad diagnostic categories of death, neoplasms and circulatory disease, for postcode sectors in the Greater Glasgow Health Board. They showed that, having allowed for the relative deprivation of each area, typical correlations between contiguous areas and those sharing common neighbours were 0.39 and 0.06 respectively for neoplasms and 0.28 and 0.19 for circulatory disease. The correlation between causes within an area was 0.28; this was 0.31 in neighbouring areas, and 0.09 in non-contiguous areas. The relatively high correlations between causes suggest that there is almost as much benefit in taking a second cause into consideration as there is in performing the spatial analysis.

Denoting the two responses by the subscripts A and B (although the method is easily generalisable to more than two responses), the logarithms of the mean number of deaths from the two causes in area i are given by

$$\left. \begin{array}{l} \log \mu_{A,i} = \log E_{A,i} + x_{0i}\beta_{A,0} + \ldots + x_{pi}\beta_{A,p} + \phi_{A,i}, \\ \log \mu_{B,i} = \log E_{B,i} + x_{0i}\beta_{B,0} + \ldots + x_{pi}\beta_{B,p} + \phi_{B,i}. \end{array} \right\} \quad (10.16)$$

The random effects for each of the heterogeneity and spatial components for each cause are given as in (10.8), but additional covariance terms are introduced such that

$$\begin{bmatrix} u_{A,i} \\ v_{A,i}^* \\ u_{B,i} \\ v_{B,i}^* \end{bmatrix} \sim N \left(\begin{bmatrix} 0 \\ 0 \\ 0 \\ 0 \end{bmatrix}, \begin{bmatrix} \sigma_{A,u}^2 & \sigma_{A,uv} & \sigma_{AB,u} & \sigma_{AB,uv} \\ \sigma_{A,uv} & \sigma_{A,v}^2 & \sigma_{AB,vu} & \sigma_{AB,v} \\ \sigma_{AB,u} & \sigma_{AB,vu} & \sigma_{B,u}^2 & \sigma_{B,uv} \\ \sigma_{AB,uv} & \sigma_{AB,v} & \sigma_{B,uv} & \sigma_{B,v}^2 \end{bmatrix} \right). \quad (10.17)$$

This permits the modelling of covariances between the heterogeneity and spatial components of the two causes.

10.5.4   Spatio-temporal models

Leyland *et al.* (1998) considered how geographically patterned observations made over time could be modelled so that more robust estimates of the relative risks for single years might be obtained through inclusion of further years' observations and in a manner such that trends over time could also be examined. The observed counts for each area were considered to be repeated measures (see Chapter 2 for a discussion of repeated measures models). The heterogeneity and spatial components for each area were assumed to follow simple linear trends; this meant that the area effect at time t, ϕ_{it}, could be written as

$$\phi_{it} = u_i + \delta_{u,i}t + \sum_j z_{ij}\left(v_j^* + \delta_{v,j}t\right), \qquad (10.18)$$

where

$$\begin{bmatrix} u_i \\ v_i^* \\ \delta_{u,i} \\ \delta_{v,i} \end{bmatrix} \sim N\left(\begin{bmatrix} 0 \\ 0 \\ 0 \\ 0 \end{bmatrix}, \begin{bmatrix} \sigma_u^2 & \sigma_{uv} & \sigma_{u\delta u} & \sigma_{u\delta v} \\ \sigma_{uv} & \sigma_v^2 & \sigma_{v\delta u} & \sigma_{v\delta v} \\ \sigma_{u\delta u} & \sigma_{v\delta u} & \sigma_{\delta u}^2 & \sigma_{\delta u\delta v} \\ \sigma_{u\delta v} & \sigma_{v\delta v} & \sigma_{\delta u\delta v} & \sigma_{\delta v}^2 \end{bmatrix}\right). \qquad (10.19)$$

This permits differential changes over time in the heterogeneity and spatial parts. Note that if $\sigma_{\delta v}^2$ is non-zero then there will be spatial patterning in the change in outcome from one year to the next.

10.6   DISCUSSION AND CONCLUSIONS

The models described in this chapter have all assumed that the disease incidence is rare and can be thought of as following a Poisson process. However, it would be a simple matter to fit a binomial model instead if this were considered appropriate (see Chapter 3 for examples of binomial regression). Other approaches have assumed that a suitable transformation of the observed and expected values may be normally distributed (see e.g. Langford et al., 1999a; Stern and Cressie, 1999); the models described in this chapter can all be adapted to the case where the underlying distribution is normal.

Section 10.5 indicated how the use of a multilevel framework can lead to a general model that can include any of the other aspects considered in this book. The introduction of higher-order autoregressive terms (or other measures of distance) is equivalent to adding a further random effect. Models with three or more levels may be of particular importance in a geographic correlation study in which the covariates may not be available at the required level of analysis. For example, a study of regional mortality may only have information on gross domestic product collected at the national level; the fact that this is a descriptor of countries rather than regions should not be ignored in the analysis. Multivariate spatial models can model the relationship between the geographical patterning of different diseases, leading, for example, to hypotheses of common aetiologies or competing causes. These models can themselves be combined; Leyland et al. (1998) give an example of a multivariate spatio-temporal model.

Although this chapter has concentrated on the use of spatial models for disease mapping and geographic correlation studies, other models could also be considered. The assessment of risk in relation to a point or line source could be modelled through the inclusion of a measure of distance of an area centroid from the putative source. Although not formalised, Section 10.4 suggests ways of looking for both the existence of clusters ($\sigma_v^2 > 0$) and the degree of clustering (through comparisons of the relative contribution of the heterogeneity and spatial components).

Estimation using iterative generalised least squares (IGLS) is fast and reasonably robust. Detailed comparisons with Bayes methods using MCMC have been conducted elsewhere (Bernardinelli and Montomoli, 1992; Mollié, 1999). The main advantage of a simulation based method such as MCMC is that it can be used to provide the full distribution of the posterior estimate of the relative risk for any particular area, whilst IGLS will produce estimates of residuals and their standard errors, the latter using sample estimates. (See the appendix to Chapter 9 for a brief introduction to MCMC.) An alternative procedure for obtaining satisfactory interval estimates is to use the bootstrap. The advantage of the methods described in this chapter is that they are computationally rapid and avoid the slow convergence problems sometimes seen in MCMC (Mollié, 1999). They can also be used to provide starting values for MCMC estimation.

CHAPTER 11

Sampling

Tom A.B. Snijders
Department of Statistics and Measurement Theory, University of Gronin-gen, The Netherlands

11.1 MULTILEVEL ANALYSIS AND MULTISTAGE SAMPLES

There is an intimate relationship between multilevel analysis and multistage samples, although this does not go as far as the two being inseparable. Multi-stage samples, as described in textbooks on sampling theory (e.g. Cochran, 1977), are useful when the population sampled is divided into subsets that may be considered exchangeable and that have some sort of administrative role. Examples are the population of inhabitants of a country divided into municip-alities, or a population of patients divided into hospitals. The subsets are conventionally called *primary sampling units* or *psu*s. In a two-stage sample, first a sample is drawn from the primary sampling units (the first-stage sample), and within each psu included in the first-stage sample, a sample of population elements is drawn (the second-stage sample). This can be extended to situations with more than two levels (e.g. individuals within households within municip-alities), and is then called a *multistage sample*. In the boundary case that each sampled psu is included entirely in the sample, i.e. the sampling fraction in the second stage is unity, the sample is called a *cluster sample*.

Clearly, multistage samples are used for precisely those nested populations where multilevel analysis can also be appropriate. The rationale, however, may be different. The usual motive for using a multistage sample is cost-efficiency: if a sample is to be drawn of 100 appendicitis patients in some year in some country, it is much cheaper to draw a two-stage sample with a first stage of, say, 10 hospitals, than to draw a simple random sample of 100 patients – who might be dispersed over 99 hospitals. On the other hand, the usual rationale for multilevel analysis resides in the research question at hand; the phenomena under study themselves have a multilevel structure, as is evident, for example, when studying contextual effects in a study of outcome measures for individuals nested in organisations (hospitals, schools, etc.) or in a longitudinal study where indi-vidual development as well as individual differences are relevant.

Multilevel Modelling of Health Statistics Edited by A.H. Leyland and H. Goldstein
©2001 John Wiley & Sons, Ltd

When a multistage sample is drawn, it usually is likely that population elements within psu's will be more alike than elements of different psu's. Some kind of multilevel analysis therefore seems called for. On the other hand, multilevel analysis can also be applied to data collected in different sampling designs. The dependence structures represented by the random intercepts and random slopes of multilevel modelling are brought about by the processes determining the phenomena under study, with or without a multistage sampling design. It can be concluded that a multistage sample will often lead to a multilevel analysis, but multilevel analysis can also be important for other data collection designs.

11.2   MODEL-BASED AND DESIGN-BASED INFERENCE

Either of two types of mechanism is usually proposed as the basis for a probability model for statistical inference. When descriptive parameters of some finite population are to be estimated from a probability sample, it is usual to base inference on the sampling design. The investigator controls the sampling process, which is the foundation for this design-based inference. An important advantage is that no extraneous assumptions are required for the unbiasedness of estimators of population parameters and the associated variance estimators.

Much statistical inference is, however, not aimed at the estimation of means or other parameters of well-defined finite populations, but rather at discovering or ascertaining mechanisms and processes in our world, reflected by the observation of measurable variables. The assumed generality of such mechanisms and processes implies that the population to which the results are supposed to apply is not only quite general but also somewhat vaguely circumscribed. Findings about the course of some disease, and the effects of relevant treatments, may be generalisable to the population of all *homines sapientes* afflicted with this disease in past, present and future – a quite hypothetical population. Results found with respect to the consequences for the course of this disease of attitudes of the patient and his or her social environment will be culture-dependent and therefore restricted to the vaguely defined population of patients living in a given culture – hypothetical, circumscribed in an unsatisfactory way, but meaningful nevertheless. Procedures of statistical inference for such investigations can be based on plausible probability models, including assertions about the distribution of random variables and their independence or conditional independence, etc. Such models do not come for free; their plausibility must be argued and their consequences checked, and if a model does not stand such tests then it must be replaced by a more plausible one. 'Random terms' or 'error terms' in such models can be regarded as resulting from influences not included among the observed variables, or – less attractively – from deviations between model and reality.

In such investigations, all or part of the sampling design often consists of just a convenience sample. In the investigation of some rare disease, the researcher will obtain collaboration from a number of clinics and include in the data all patients in these clinics suffering from this disease. The results of the investigation may be thought to apply to anyone suffering from this disease. To argue that the patients included in the study can be considered a random sample from this population, the investigator has to consider carefully the selection processes that lead to a patient being included in the study, and whether there could be unobserved factors to do with severity of the disease, comorbidity, general health status, etc., that are related simultaneously to the selection of the patient in the study and to the measured variables. Only if it is plausible that no such variables exist is it reasonable to apply model-based procedures of statistical inference. Often such considerations lead to circumscribing the population to which the results can be generalised; for example, patients who have been suffering from the disease for a protracted period or those who are well motivated to comply with their therapy.

The multilevel statistical procedures treated in this book are examples of model-based statistical inference. For example, the usual two-level hierarchical linear model implies assumptions of independence between level-2 units, conditional independence between level-1 units within each level-2 unit given the random effects associated with this level-2 unit, and normal distributions for the error terms. The investigator must check critically whether these assumptions are plausible. If they are, the data can be analysed as if they are produced by a two-stage sample with random selection in both stages, although it is not necessary that the sampling procedure actually was carried out in this way.

If, on the other hand, one wishes to follow a design-based approach – for example because the study has a descriptive purpose – and the selection probabilities are not constant, the sampling design must be taken into account to obtain unbiased estimators. For the estimation of population means, this is treated in the standard textbooks about sampling theory that include multistage sampling (e.g Cochran, 1977). For more general statistical questions, such as hypothesis testing and regression analysis, this is treated in specialised literature (e.g. Skinner *et al.*, 1989). For the estimation of parameters in the hierarchical linear model, however, it is much more complicated to take unequal selection probabilities into account. Methods to do so are proposed in Pfefferman *et al.* (1998). The remainder of this chapter is about design-based inference only.

11.3   STUDY DESIGN: POWER AND STANDARD ERRORS

The following sections of this chapter are mainly about the design of two-level studies and, in particular, the determination of optimal or adequate sample sizes. What complicates the choice of an adequate design for a multilevel study is the fact that there are sample sizes to be chosen at each level of the nesting hierarchy. For example, when studying patients in hospitals, the researcher has to decide whether a given number of hospitals and a given number of patients within each hospital is adequate. In another example, when some outcome variable is measured repeatedly for a sample of patients, it has to be decided how many patients to include in the study and how often to measure the outcome variable for each of them. Another important choice in multilevel experimental design that does not occur in single-level designs is the determination of the *level* of randomisation. For example, when studying a new medical treatment, the researcher may have to choose between randomising within and randomising between hospitals.

Since considerations for the choice of a design are always of an approximate nature, only balanced designs are considered here, i.e. those designs where each level-2 unit contains the same number of level-1 units. Level-2 units will sometimes be referred to as *clusters*. The number of level-2 units is denoted by m, and the number of level-1 units within each level-2 unit is denoted by n. These numbers are called the *level-2 sample size* and the *cluster size* respectively. The total sample size is $N = mn$. If, in reality, the number of level-1 units fluctuates between level-2 units, it will almost always be a reasonable approximation for n to use the average number of sampled level-1 units per level-2 unit.

Optimality or adequacy of the design is primarily a function of the power of tests and the standard errors of estimators. This chapter concentrates on parameters in the fixed part of the model for which the estimator is approximately normally distributed. Denote this parameter by β and the standard error of estimation by SE $(\hat{\beta})$. Provided that sample sizes are not very small, the test for β can be approximated by the standard normal test applied to the t-ratio $\hat{\beta}/\mathrm{SE}(\hat{\beta})$. If the significance level is denoted by α and the power by γ, the approximate relation between standard error and power is

$$\frac{\beta}{\mathrm{SE}(\hat{\beta})} \approx (z_{1-\alpha} + z_\gamma) = (z_{1-\alpha} - z_{1-\gamma}), \tag{11.1}$$

where $z_{1-\alpha}$, z_γ and $z_{1-\gamma}$ are the values for which the standard normal distribution has the indicated cumulative probability values. For example, if $\alpha = 0.05$ and a power is desired of $\gamma = 0.80$ if the effect size is $\beta = 0.20$ then the standard error should be no more than

$$\text{standard error} \leq \frac{0.20}{1.64 + 0.84} = 0.081.$$

It should be stressed that this approximation does not take into account the degrees of freedom for variance estimation, and it might be relevant to modify the conclusions of the following analysis for relatively low values of m and n. In the following sections, the discussion will be mainly in terms of standard errors.

11.4 THE DESIGN EFFECT FOR THE ESTIMATION OF A MEAN

To discuss the estimation of fixed effect parameters, first three important special cases are considered: the estimation of a grand mean, the estimation of the regression coefficient of a level-2 variable, and the estimation of such a coefficient of a level-1 variable without any level-2 variance. This should give the reader an understanding of some issues which are important for standard errors of estimators for such parameters. Following this the general case will be discussed.

The estimation of a population mean is a scientific question where model-based and design-based inference meet. We approach it in a model-based way, but a design-based approach for sampling from a finite population arrives at basically the same answers, if the sample is a two-stage sample using random sampling with replacement at either stage or if the sampling fractions are so low that the difference between sampling with and sampling without replacement is negligible.

Suppose that the mean is to be estimated of some variable Y in a population that has a two-level structure. As an example, Y could be the duration of hospital stay after a certain operation under the condition that there are no complications or additional health problems. Suppose also that it is reasonable to postulate the empty model of multilevel analysis,

$$y_{ij} = \beta_0 + u_j + e_{ij},$$

with the usual assumptions. The variance of the random intercept is $\mathrm{var}(u_j) = \sigma_{u0}^2$ and the level-1 variance is $\mathrm{var}(e_{ij}) = \sigma_e^2$. The parameter to be estimated is β_0. The overall sample mean

$$\hat{\beta}_0 = \frac{1}{mn} \sum_{j=1}^{m} \sum_{i=1}^{n} y_{ij}$$

is the obvious estimator, and its variance is

$$\mathrm{var}\left(\hat{\beta}_0\right) = \frac{n\sigma_{u0}^2 + \sigma_e^2}{mn}.$$

The sample mean of a simple random sample of N elements from this population has variance

$$\frac{\sigma_{u0}^2 + \sigma_e^2}{mn}.$$

The relative efficiency of the simple random sample with respect to the two-stage sample is the ratio of these variances:

$$\frac{n\sigma_{u0}^2 + \sigma_e^2}{\sigma_{u0}^2 + \sigma_e^2} = 1 + (n-1)\rho_I, \qquad (11.2)$$

where ρ_I is the intraclass correlation coefficient,

$$\rho_I = \frac{\sigma_{u0}^2}{\sigma_{u0}^2 + \sigma_e^2}.$$

The quantity (11.2) is called the *design effect* of the two-stage sample (see e.g. Cochran, 1977). It is the ratio of the variance obtained with the two-stage sample to the variance obtained for a simple random sample with the same total sample size. A large design effect means statistical inefficiency, but this disadvantage may be offset by the cost reductions of the two-stage design. A two-stage sample yields the same standard error as a simple random sample for which the total sample size is divided by the factor in (11.2).

11.5 THE EFFECT OF A LEVEL-2 VARIABLE

When a two-level regression is carried out and the objective is to estimate the regression coefficient of a level-2 variable X, the estimated coefficient is practically equivalent to the estimated regression coefficient in the single-level regression analysis for data aggregated to the cluster means of all relevant variables. If the variance of the random intercept is again denoted by σ_{u0}^2 and the residual level-1 variance by σ_e^2, the residual variance for the aggregated regression analysis is $\sigma_{u0}^2 + \sigma_e^2/n$. Assume that Y is distributed according to a random intercept model:

$$y_{ij} = \beta_0 + \beta_1 x_j + u_j + e_{ij}.$$

When the variable X has variance s_x^2, the variance of the estimated regression coefficient is

$$\mathrm{var}\left(\hat{\beta}_1\right) = \frac{n\sigma_{u0}^2 + \sigma_e^2}{mns_X^2}. \tag{11.3}$$

This implies that again the relative efficiency of the two-stage sampling design is given by (11.2).

Thus it appears that for estimating a population mean or, more generally, the effect of a level-2 variable, and if the intraclass correlation is moderate or high, a large cluster size leads to a large statistical inefficiency in estimating the population mean.

11.6 THE EFFECT OF A LEVEL-1 VARIABLE

Now consider the opposite situation, where one wishes to estimate the regression coefficient of an independent variable X that is a pure level-1 variable; that is, its mean is the same in each level-2 unit. Note that this implies that the intraclass correlation of X is negative, $-1/(n-1)$, which implies that X itself is not distributed according to the hierarchical linear model (which allows only non-negative intraclass correlations). For simplicity, assume that the cluster mean of X is 0 and its variance is the same within each cluster, denoted by s_X^2. An example is the effect of a treatment that is randomly allocated to a fixed fraction of the level-1 units within each level-2 unit. Another example is the linear effect of time in a balanced longitudinal design.

The estimator for the regression coefficient is now the average of the within-cluster regression coefficients:

$$\hat{\beta}_1 = \frac{1}{mns_X^2} \sum_{j=1}^{m} \sum_{i=1}^{n} x_{ij} y_{ij}. \tag{11.4}$$

If Y is distributed according to a random intercept model,

$$y_{ij} = \beta_0 + \beta_1 x_{ij} + u_j + e_{ij},$$

then the assumptions about the variable X imply that the estimator (11.4) is equal to

$$\hat{\beta}_1 = \beta_1 + \frac{1}{mns_X^2} \sum_{j=1}^{m} \sum_{i=1}^{n} x_{ij} e_{ij}, \tag{11.5}$$

and its variance is

$$\operatorname{var}\left(\hat{\beta}_1\right) = \frac{\sigma_e^2}{mns_X^2}. \tag{11.6}$$

A simple random sample of N elements from the same distribution yields an OLS estimator for β_1 with variance

$$\frac{\sigma_{u0}^2 + \sigma_e^2}{mns_X^2}$$

(assuming that the variance of X in this sample is also precisely s_X^2). This shows that the two-stage sample here is more efficient than the simple random sample, with the design effect

$$\frac{\sigma_e^2}{\sigma_{u0}^2 + \sigma_e^2} = 1 - \rho_I. \tag{11.7}$$

The greater efficiency in this case of the two-stage sample is well known in experimental design: the two-stage design corresponds to blocking on the level-2 units. In psychology this is the frequently used within-subject design. Blocking is known to neutralise the main block effect as a variance component.

Since the design effect is less than unity for level-1 variables and larger than unity for level-2 variables, it may be concluded that if the study is a comparison of randomly assigned treatments in a random intercept model and the study costs are determined by the total sample size N then randomising within clusters is more efficient than randomising between clusters. The optimal level of randomisation for two- and three-level designs is discussed extensively by Moerbeek *et al.* (2000).

Example Suppose that a new training program ('treatment') for nurses is to be compared with an existing training program ('control'), while hospitals are believed to be a major influence on the nurses' work. The research question can be phrased in terms of the preceding sections as the estimation of the regression coefficient of the dummy variable that distinguishes treatment from control. Denote this variable by X, defined as 0 for the control and 1 for the treatment condition. When the treatment fraction is p, its variance is $\text{var}(X) = p(1 - p)$; for $p = 0.5$, this yields $s_X^2 = 0.25$. Assume that the dependent variable is standardised to have a unit variance, and that it has an intraclass correlation of 0.10. Further assume that the treatment is equally effective for all hospitals, i.e. the random intercept model is adequate. Then $\sigma_e^2 = 0.9$ and $\sigma_{u0}^2 = 0.1$. Then the estimation variance for level-2 randomisation is (11.3)

$$\frac{0.4}{m} + \frac{3.6}{mn},$$

and for level-1 randomisation it is (11.6)

$$\frac{3.6}{mn}.$$

If group sizes n are predetermined, the advantage of randomisation at level 1 is quite large.

11.6.1   A level-1 variable with a random slope

The level-1 variable , however, may well have a random slope in addition to the random intercept. The model for Y then reads

$$y_{ij} = \beta_0 + \beta_1 x_{ij} + u_{0j} + u_{1j}x_{ij} + e_{ij}. \tag{11.8}$$

Denote the intercept variance by σ_{u0}^2 and the slope variance by σ_{u1}^2. For the random slope model, the variance of the estimated regression coefficient (11.4) is

$$\text{var}\left(\hat{\beta}_1\right) = \frac{n\sigma_{u1}^2 s_X^2 + \sigma_e^2}{mns_X^2}. \tag{11.9}$$

The total residual variance of Y (where the level-2 unit is random, i.e. marginalised) is equal to

$$\sigma_e^2 + \sigma_{u0}^2 + \sigma_{u1}^2 s_X^2,$$

so that the design effect is now

$$\frac{n\sigma_{u1}^2 s_X^2 + \sigma_e^2}{\sigma_{u0}^2 + \sigma_{u1}^2 s_X^2 + \sigma_e^2}.$$

This shows that the two-stage sample with level-1 randomisation only 'neutralises' the random intercept and not the random slope of X as terms in the variance of the estimated regression coefficient.

   In practice, the presence of a random slope for the variable X means that the regression coefficient β_1 does not tell all of the story, and it is important to estimate the random slope variance σ_{u1}^2 as well. This underscores the fact that design considerations should never focus narrowly on the estimation of just one statistical parameter.

11.7   OPTIMAL SAMPLE SIZE FOR ESTIMATING A REGRESSION COEFFICIENT

In studies leading to statistical models such as those treated in the preceding sections, the design is determined by the sample sizes m and n and the distribution of the X-values. This distribution has a within-cluster and a between-cluster aspect. As in OLS regression, if one has the liberty to choose the X-values, it is optimal to maximise their dispersion. With respect to optimal sample sizes, the multilevel or two-stage design requires the determination of the sample sizes at the two levels. This section is about the optimal choice of these sample sizes for the estimation of a regression coefficient under given budget constraints.

If the aim is to have a minimum variance for a given total sample size N and a given value for s_X^2 then it is clear from (11.6) that for a within-cluster deviation variable without level-2 slope variation, it does not matter how the total sample size is distributed over the level-2 units, as long as one succeeds in constructing an X-variable with constant within-cluster mean and within-cluster variance s_X^2. This implies, of course, that n is at least 2. For a level-2 variable or a within-cluster deviation variable with positive random slope variation, (11.3) and (11.9) imply that it is optimal to let m be as large as possible. This would imply $n = 1$, i.e. a simple random sample is optimal. If this is not feasible then still it is best to have the clusters as small as possible.

Usually, however, study costs are not a function of total sample size but depend on both the total sample size and the number of level-2 units. The costs often are well approximated by a function of the type $c_1 m + c_2 mn$. Thus an optimal design is obtained when the variance of the estimator is minimal, given the constraint

$$c_1 m + c_2 mn \leq k, \tag{11.11}$$

where k denotes the total budget. In the preceding sections, it was shown that the variance to be minimised can be expressed as

$$\frac{\sigma_1^2}{m} + \frac{\sigma_2^2}{mn},$$

for a suitable choice of σ_1^2 and σ_2^2, which can be found in equation (11.3), (11.6) or (11.9). The minimisation of this expression under the constraint $c_1 m + c_2 mn = k$ is treated by Cochran (1977, Section 10.6). The optimal value for n is

$$n_{opt} = \sqrt{\frac{c_1 \sigma_2^2}{c_2 \sigma_1^2}}, \tag{11.12}$$

rounded upwards or downwards to an integer value. This is also the optimal n if the budget is to be minimised under the constraint that the estimation variance has a preassigned value. It may be noted that the optimal cluster size does not depend on the available budget or on the level-2 sample size.

For explanatory variables defined at level 1 and having a constant mean across level-2 units, and for which the dependent variable follows a random intercept model, it can be seen from (11.9) that $\sigma_1^2 = 0$, so that n_{opt} is infinite. This means in practice that n should be as large as possible: a single-level design, for which $m = 1$ and n is the total sample size, is preferable to a two-level design for the estimation of a regression coefficient of a level-1 variable when the budget constraint is given by (11.11). If the explanatory variable X is defined at level 2, on the other hand, we have $\sigma_1^2 = \sigma_{u0}^2/s_X^2$ and $\sigma_2^2 = \sigma_e^2/s_X^2$, so that the optimal cluster size is

$$n_{opt} = \sqrt{\frac{c_1 \sigma_e^2}{c_2 \sigma_{u0}^2}}.$$

This optimal sample size is also discussed by Raudenbush (1997, p. 177) and Moerbeek *et al.* (2000). The latter paper also treats optimal allocation for three-level designs. Optimal allocation for two and three-level designs for binary responses are discussed by Moerbeek *et al.* (2001).

11.8   THE USE OF COVARIATES

It is well known in experimental design that controlling for relevant covariates can lead to important gains in efficiency. In a single-level design, a covariate that has a residual correlation with the dependent variable equal to ρ will yield a reduction of the unexplained variance by a factor $1 - \rho^2$. When the sample size is large enough for the loss of a degree of freedom for the variance estimate to be unimportant, this will allow the researcher to diminish the sample size by this factor while retaining the same standard error and power.

For a two-level design, the situation is – of course – more complicated. The reduction in standard error depends on the intraclass correlation of the dependent variable and on the within-group and the between-groups residual correlations between the dependent variable and the covariate. For more precise calculations in small-sample situations, the degrees of freedom also play a part, but this is not considered in the following analysis.

Suppose that, as above, we wish to analyse the regression coefficient of some variable X, and we are interested to see how much gain in precision is obtained by controlling for some covariate denoted by Z. The model without control for Z is supposed to be the random intercept model

$$y_{ij} = \beta_0 + \beta_1 x_{ij} + u_j + e_{ij},$$

while the model with control for Z is

$$y_{ij} = \tilde{\beta}_0 + \beta_1 x_{ij} + \beta_2 z_{ij} + \tilde{u}_j + \tilde{e}_{ij}.$$

It is assumed that Z follows a random intercept model:

$$z_{ij} = \gamma_0 + u_{zj} + e_{zij}.$$

Further assume that Z is, both within and between groups, uncorrelated with X. Then the regression coefficient β_1 remains the same when controlling for Z, which is reflected in the notation used in the preceding equations.

Denote the population residual within-group correlation between Y and Z by ρ_W and the population residual between-group correlation by ρ_B. These are defined by (cf. Snijders and Bosker, 1999, Section 3.6)

$$\rho_W = \rho(e_{ij}, e_{zij}), \qquad \rho_B = \rho(u_j, u_{zj}).$$

Some calculations show that the reduction in the variance parameters due to the control for Z is given by

$$\tilde{\sigma}_e^2 = (1 - \rho_W^2)\sigma_e^2, \quad \tilde{\sigma}_{u0}^2 = (1 - \rho_B^2)\sigma_{u0}^2. \tag{11.13}$$

In a large-sample approximation (valid when n and m are large), these reductions are applied to the estimation variances given in (11.3) and (11.6). Thus, if X is a level-2 variable then formula (11.3) applies and the factor $\sigma_{u0}^2 + \sigma_e^2/n$ will be replaced by $(1 - \rho_B^2)\sigma_{u0}^2 + (1 - \rho_W^2)\sigma_e^2/n$. If X is a level-1 variable then the factor σ_e^2 in (11.6) is replaced by $(1 - \rho_W^2)\sigma_e^2$. This illustrates that for pure level-1 variables, it is – naturally – only the within-group correlation that counts, whereas for level-2 variables, not only the between-group correlation but also the within-group correlation plays a role in the reduction of the estimation variance of β_1. If group sizes n are large, however, the influence of the within-group correlation for level-2 variables will be of minor importance.

When one investigates the effect of a level-2 variable X controlling for a level-1 variable Z, one may be tempted to use the group means of Z rather than their individual values. The preceding analysis demonstrates that this leads to a loss in estimation efficiency. If n is large, the loss will be negligible. This point is also made by Raudenbush (1997), who gives a more extensive discussion of the use of covariates, taking into account also the random nature of the observed residual covariances between Z and Y (but not the loss in degrees of freedom).

11.9 STANDARD ERRORS FOR FIXED EFFECTS IN GENERAL

In practice, the assumptions made in Sections 11.4 and 11.5 often are not an adequate simplification of reality; moreover, many researchers wish to estimate several regression coefficients from a single data set as precisely as possible. Exact formulae for estimation variances are not available for arbitrary multilevel designs. Snijders and Bosker (1993) derived approximate formulae for estimation variances in two-level designs, valid under the restriction that variables with random slopes have a zero between-cluster variance and that n is not too small – say at least 8. These formulae are calculated using the computer program PinT ('Power in Two-level designs'), which can be downloaded, with manual, from

http://stat.gamma.rug.nl/snijders/multilevel.htm

The main difficulty in applying this program is the requirement that plausible parameter values be specified. This is, of course, a general difficulty in any power analysis (cf. Kraemer and Thiemann, 1987; Cohen, 1992), but it is more pressing in the case of multilevel analysis because the random part parameters must also be specified. The use of PinT will be illustrated here by means of an example. The PinT manual and Chapter 10 of Snijders and Bosker (1999) contain various other examples.

As an example, suppose that one is investigating the effect of the training of psychotherapists on therapy effectivity. Level-1 units are patients and level-2 units are therapists. The dependent variable is a patient-level outcome measure standardized to unit variance. The investigated training is a course represented by a 0–1 variable. In addition, the level of professional training of the psychotherapists and a pretest measure of the seriousness of the patients' complaints are relevant. The question is: how many patients and how many therapists have to be investigated?

Assume that participation in the course will be randomised within groups of equal professional training in such a way that the fraction following the course is higher in the groups with lower professional training. Professional training and pretest are represented by numerical variables also standardised to mean 0 and unit variance. If a random intercept model applies, with these three variables having fixed effects, the model can be expressed as

$$y_{ij} = \beta_0 + \beta_1 x_{1j} + \beta_2 x_{2j} + \beta_3 x_{3ij} + u_j + e_{ij},$$

where X_1 indicates whether the therapist followed the course, X_2 is therapists' professional training and X_3 is the pretest. The primary research variable is X_1.

The required information for running PinT consists of the means, variances and covariances of the explanatory variables and all parameters of the random part.

First consider the means. The variables X_2 and X_3 have means 0. Suppose that the fraction of therapists who follow the course is thought to be 0.4. This is the mean of X_1.

Now consider the variances and covariances of the explanatory variables. Suppose that the therapists are somewhat different in the pretest values of their patients, this variable having an intraclass correlation of 0.19. The variance of X_1, being a binary variable with mean 0.4, is 0.24. Suppose that the correlation between the therapist mean of the pretest and the professional training X_2 is known to be 0.5. Assume that the randomisation of the course participation, which is conditional on X_2, will give a correlation between X_1 and X_2 of -0.4. The partial correlation between pretest mean and course participation, controlling for level of professional training, is expected to be zero, which leads to a total correlation between pretest mean and course participation of 0.2. The within-groups variance of X_3 is then $\sigma^2_{X(W)} = 1 - 0.19 = 0.81$, and the between-groups covariance matrix of (X_1, X_2, X_3) is

$$\Sigma_{X(B)} = \begin{pmatrix} 0.24 & -0.20 & 0.043 \\ -0.20 & 1.00 & 0.22 \\ 0.043 & 0.22 & 0.19 \end{pmatrix}.$$

Finally consider the parameters of the random part of the multilevel model. To get some insight into plausible values of the level-1 and level-2 variances, it may be helpful to note that the variance of the dependent variable can be decomposed as

$$\text{var}(Y_{ij}) = \beta_3^2 \sigma_{X(W)}^2 + \boldsymbol{\beta}^{\mathsf{T}} \Sigma_{X(B)} \boldsymbol{\beta} + \sigma_{u0}^2 + \sigma_e^2,$$

where $\boldsymbol{\beta} = (\beta_1, \beta_2, \beta_3)^{\mathsf{T}}$ (cf. Snijders and Bosker, 1999, Section 7.2). This corresponds, for the variance decomposition of Y_{ij} in the empty model, to a total level-1 variance of $\beta_3^2 \sigma_{X(W)}^2 + \sigma_e^2$ and a total level-2 variance of $\boldsymbol{\beta}^{\mathsf{T}} \Sigma_{X(B)} \boldsymbol{\beta} + \sigma_{u0}^2$.

Assume that the total level-1 and level-2 variances of the outcome measure are 0.8 and 0.2 respectively, and that the available explanatory variables together explain 0.25 of the level-1 variance and 0.5 of the level-2 variance. Then $\sigma_e^2 = 0.6$ and $\sigma_{u0}^2 = 0.10$. In terms of the decomposition of total variance, this corresponds to a raw explained level-one variance of $\beta_3^2 \sigma_{X(W)}^2 = 0.2$, and therefore a regression coefficient $\beta_3 = \sqrt{0.20/0.81} = 0.5$, and $\boldsymbol{\beta}^{\mathsf{T}} \Sigma_{X(B)} \boldsymbol{\beta} = 0.1$.

With respect to the cost structure, assume that the budget constraint can be expressed as (11.11) with $c_1 = 20$, $c_2 = 1$ and $k = 1000$. In other words, an extra therapist in the sample costs 20 times as much as an extra patient; and there would be enough funds to include, for example, 40 therapists with 5 patients each, or 20 therapists with 30 patients each.

With this specification, PinT can be executed, and it produces the approximate standard errors of the estimated fixed effects for sample sizes satisfying the budget constraint $20m + mn \le 1000$. The standard errors for β_1 are plotted in Figure 11.1. The plot is rather irregular owing to the inequality constraint for

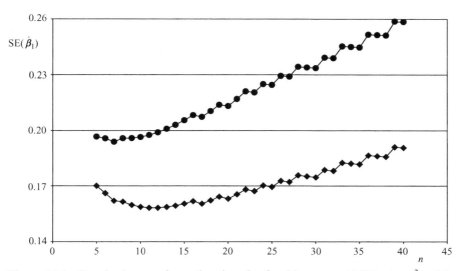

Figure 11.1 Standard error for estimating β_1, for $20m + mn \le 1000$: $\blacklozenge$, $\sigma_e^2 = 0.6$, $\sigma_u^2 = 0.1$; $\bullet$, $\sigma_e^2 = 0.5$, $\sigma_u^2 = 0.2$.

the integer numbers m and n. The minimum is seen to be rather flat. The minimum standard error is 0.156, achieved for $n = 11$. For cluster sizes between 7 and 18, the standard error is less than 0.162. It can be concluded that, for these parameter values, average cluster sizes between 7 and 18 are fully acceptable.

To investigate the sensitivity of this result to the assumed parameter values, the calculations were also carried out for $\sigma_e^2 = 0.5$ and $\sigma_{u0}^2 = 0.2$, with all other parameters remaining the same. These results are also shown in Figure 11.1. The greater level-2 variance leads to a bigger standard error for this level-2 variable. The minimum is 0.192 for $n = 7$. The minimum is less flat than for the earlier parameter values; for $n \leq 13$, the standard error is less than 0.200. For these parameter values, the average cluster sizes would preferably be 13 or less. Note, however, that in the second situation, the intraclass correlation is twice as big as in the first one, so the two situations are quite different.

The PinT program uses a rather rough large-sample approximation to obtain the standard errors. This is often adequate, because design questions are usually of a very approximate nature, but more precise approximations are desirable when they are available. For the special case of testing the effect of a level-2 variable (representing the difference between a treatment and a control condition in a cluster randomised trial), controlling for a single level-1 covariate, a more precise approximation is given by Raudenbush (1997, pp. 178–179). He obtains a result that, for larger sample sizes, boils down to (11.12). The greater precision can be important for small sample sizes.

11.10 RANDOM PARAMETERS

Usually the focus of research questions is on the fixed parameters. Sometimes, however, the design should also be adequate for the estimation of the random parameters. This section considers the estimation of the intraclass correlation coefficient. For some remarks about the design of multilevel studies with respect to the estimation of other random parameters, see Mok (1995), Cohen (1998), and Snijders and Bosker (1999, Section 10.5.2).

Donner (1986) proved that the standard error of the estimated intraclass correlation coefficient in an empty two-level model (i.e. a two-level model without any explanatory variables) with constant cluster size n is given by

$$\text{SE}(\hat{\rho}_I) = (1 - \rho_I)[1 + (n - 1)\rho_I]\sqrt{\frac{1}{n(n-1)(m-1)}}. \tag{11.14}$$

This standard error depends on the parameter itself that is to be estimated. To obtain optimal sample sizes for estimating the intraclass correlation given the budget constraint (11.11), it is convenient to substitute $m = k/(c_1 + c_2 n)$, which transforms (11.14) into a function of n, which can be plotted. From the graph, the optimum value for n can be deduced, as well as the sensitivity of this minimum to suboptimal values of n.

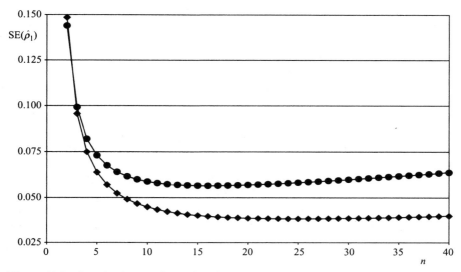

Figure 11.2 Standard error for estimating the intraclass correlation coefficient for a budget constraint $20m + mn \leq 1000$ with $\rho_I = 0.1$ and 0.2: ◆, $\rho_I = 0.1$; ●, $\rho_I = 0.2$.

As an example, suppose that it is desired to estimate the intraclass correlation with a budget constraint $20m + mn \leq 1000$ and that the intraclass correlation is believed to be between 0.1 and 0.2. Figure 11.2 shows the standard errors of the intraclass correlation coefficient, using the substitution $m = 1000/(20 + n)$ (ignoring the integer nature of the sample sizes), for and $\rho_I = 0.1$ and 0.2. For $\rho_I = 0.1$, the minimum standard error is 0.03835, achieved for $n = 24$ and 25, and the standard error is less than 0.040 for n between 16 and 40. For $\rho_I = 0.2$, the minimum standard error is 0.05645, achieved for $n = 16$, the standard error being less than 0.059 for n between 10 and 27. In order to have a relatively small standard error for ρ_I in the range between 0.1 and 0.2, cluster sizes between 16 and 27 are fully acceptable.

11.11   CONCLUSIONS

Important statistical issues in designing a powerful multilevel study are the choice of covariates, the level of randomization, and the determination of sample sizes at the various levels. This chapter has made some remarks about the first two, and has concentrated on the last issue. Usually, the determination of sample sizes in a multilevel investigation is at best a matter of well-informed and well-educated guesswork. In the first place, one has to choose the main statistical parameters for which an adequate (or even optimal) design is desired. In some focused experiments, this is the regression effect of a treatment variable. In many other situations, a more extensive number of parameters – primarily regression coefficients but possibly including the intraclass correlation and variance components – are of major interest. The design then has to ensure that the standard errors for these main parameters are small enough.

To make an a priori assessment of standard errors of estimation for various parameters, one has to determine, or guess, the values of parameters such as variances and covariances of outcome and explanatory variables for each relevant level in the design. This was illustrated above by some examples. Sometimes such guesses can reasonably be made on the basis of existing data; otherwise, it is important to conduct a sensitivity analysis by varying the guessed parameter values and studying how this affects the standard errors of interest.

Given the complexity of design considerations for multilevel studies, it is often advisable to reduce the problem to its simplest form, ignoring control variables for which a minor impact is expected, and start with a random intercept model. If there are more than two levels involved and it is possible to point out the higher level that is expected to be associated with the largest random variability, then it may be advisable to ignore temporarily the other levels and start with a power analysis for a two-level model. If such a simplified first analysis has provided some rough insight into the effects of various combinations of sample sizes at the various levels on standard errors and/or power, and if there is sufficient information to make guesses about the values of additional parameters, then in a further analysis one could enter a random slope in the design considerations, and perhaps more than two levels. The general rule is to start design considerations as simply as possible, because if one tries to face the complexity right away one runs the risk of being put off and cancelling the a priori design considerations altogether – and wouldn't that be a pity?

CHAPTER 12

Further Topics in Multilevel Modelling

Harvey Goldstein
Mathematical Sciences, Institute of Education, University of London, UK

Alastair H. Leyland
MRC Social and Public Health Sciences Unit, University of Glasgow, UK

12.1 INTRODUCTION

This chapter covers further topics in multilevel modelling that will be of particular interest to those involved in research in the health sciences. The following sections therefore illustrate how the theory and methods underlying multilevel models may be extended to include meta-analysis, survival data modelling, and the ideas of context and composition. This is followed by a brief discussion of recent developments in multilevel modelling.

12.2 META-ANALYSIS

The purpose of meta-analysis is to provide an overall summary of results when information from several studies of the same topic are available. These 'studies' may be centres in a single clinical trial, distinct experimental studies, distinct (or possibly overlapping) observational surveys, or mixtures of these. Meta-analysis can therefore be regarded as a special case of the general hierarchical data model, where individual observations are nested within studies or centres. Viewing meta-analysis within this framework leads to some important and natural extensions.

In applied work, it is often assumed that the effect of interest is constant across the component studies (Thompson and Pocock, 1991), yielding the so-called 'fixed effect' model. The assumption of homogeneity can, however, be relaxed to allow for random variation between studies of the effects, yielding the so-called 'random effects' model (DerSimonian and Laird, 1986). Statistical

Multilevel Modelling of Health Statistics Edited by A.H. Leyland and H. Goldstein
©2001 John Wiley & Sons, Ltd

models for this case can be fitted using a variance components multilevel model formulation. A general multilevel formulation (Goldstein, 1995), however, allows more general random coefficient models to be studied, and we describe this in more detail below. A straightforward extension is to include covariates in such a model and to observe the extent to which they account for between-study variation. An additional problem is when some studies provide individual-level data, while for others only summary results (such as means) are available and methods of meta-analysis that can combine such results efficiently are now available (Goldstein *et al.*, 2000).

For aggregate-level data, consider the following underlying model for individual-level data, for example a measure of attitude towards health education in schools where we have pupils grouped within studies with a treatment group that has been exposed to health education and a control group that has not. Suppose that we have a basic model, with the response Y_{ij} being the attitude score (suitably transformed to normality) for the ith pupil in the jth study, as

$$\left.\begin{array}{l} y_{h,ij} = \beta_0 + \beta_1 x_{ij} + \beta_2 t_{h,ij} + u_{h,j} + e_{h,ij}, \\ \mathrm{var}(u_{h,j}) = \sigma_{hu}^2, \qquad \mathrm{var}(e_{h,ij}) = \sigma_{he}^2, \end{array}\right\} \tag{12.1}$$

with the usual assumptions of normality and independence. The term x_{ij} is a covariate, in this case a baseline pretreatment measure of attitude. The subscript h indexes the treatment/control and the term $t_{h,ij}$ is 1 if treatment and 0 if control. The random effect $u_{h,j}$ is a study effect and the $e_{h,ij}$ are individual-level residuals. Clearly this model can be elaborated in a number of ways, by including further covariates at study or individual level, by allowing β_2 (or β_1) to vary at level 2 so that the effect of treatment varies across studies, and by allowing the level-1 variance to depend on other factors such as gender or ethnic origin. These generalisations are discussed in Goldstein *et al.* (2000).

Suppose now that we do not have individual data available but only means at the study level. If we average (12.1) to the study level, we obtain

$$y_{h,j} = \beta_0 + \beta_1 x_{.j} + \beta_2 t_{h,j} + u_{h,j} + e_{h,.j}, \tag{12.2}$$

where $y_{h,j}$ is the mean response for the jth study for treatment/control (h). The residual variance for this model is given by

$$\sigma_{hu}^2 + \sigma_{he}^2 / n_{hj},$$

where n_{hj} is the number of pupils in treatment h for the jth study. It is worth noting at this point that we are ignoring, for simplicity, levels of variation within studies, which will add further levels to the model. If we have information on the relevant quantities in (12.2) then we shall be able to obtain estimates for the model parameters, so long as the n_{hj} differ. Such estimates, however, may not be very precise and extra information, especially about the value of the level-1 variances, will improve them.

Model (12.2) forms the basis for the multilevel modelling of aggregate-level data. In practice, the results of studies will often be reported in non-standard form, for example with no estimate of σ_{he}^2, but it may be possible to estimate this from reported test statistics. In some cases, however, the reporting may be such that the study cannot be incorporated in a model such as (12.2). Goldstein *et al.* (2000) give a set of minimum reporting standards in order that meta-analysis can subsequently be carried out.

12.2.1 Combining individual-level data with aggregate-level data

While it is possible to perform a meta-analysis with only aggregate-level data, it is clearly more efficient to utilise individual-level data where these are available. In general, therefore, we shall need to consider models that have mixtures of individual and aggregate data, even perhaps within the same study.

We can do this by specifying a model that is just the combination of (12.1) and (12.2), namely

$$\left.\begin{array}{l} y_{h,ij} = \beta_0 + \beta_1 x_{ij} + \beta_2 t_{h,ij} + u_{h,j} + e_{h,ij}, \\ y_{h,j} = \beta_0 + \beta_1 x_j + \beta_2 t_{h,j} + u_{h,j} + e_{h,j} z_{h,j}, \\ z_{h,j} = \sqrt{n_{hj}^{-1}}, \quad e_{h,j} \equiv e_{h,ij}. \end{array}\right\} \qquad (12.3)$$

What we see is that the common level-1 and level-2 random terms link together the separate models and allow a joint analysis that makes fully efficient use of the data. Several issues immediately arise from (12.3). One is that the same covariates should be involved. This is also a requirement for the separate models. If some covariate values are missing at either level then it is possible to use an imputation technique to obtain estimates, assuming a suitable random missingness mechanism. The paper by Goldstein *et al.* (2000) discusses generalisations of (12.3) and applies it to an analysis of class size studies.

12.2.2 Clinical trial meta-analysis

One of the most common applications of meta-analysis in medicine is to clinical trials with a basic binary response. This involves a series of choices. The decisions at each stage are similar whether the meta-analyst has only summary data from published results or full individual patient data, but the options available may differ. The first choice is between the fixed effect and random effects models, and in either case the method of estimation must be selected from a number of alternatives. If one is fitting a random effects model, more decisions arise: how to allow for uncertainty in estimation of the between trial variance when constructing a confidence interval for the treatment effect, how to obtain confidence intervals for the between-trial variance, how to incorporate trial-level covariates and how to investigate between-trial heterogeneity.

The usual fixed effect model for meta-analysis assumes the true treatment effects to be homogeneous across trials, and accordingly estimates the common treatment effect θ by a weighted average of the trial-specific estimates, with weights equal to the reciprocals of their within-trial variances. The random effects two-level model assumes that the true treatment effects vary randomly between trials. This model therefore includes a between-trial component of variance, say τ^2. A commonly used measure of treatment effect in binary event data is the log odds ratio; the normality assumption required is more easily satisfied for this than for alternative measures such as the risk difference. We write a model analogous to (12.2) as

$$
\begin{aligned}
y_j &= \theta + v_j + e_{.j}, \\
v_j &\sim N(0, \tau^2), \quad \text{var}(e_{.j}) = \sigma_{ej}^2,
\end{aligned}
\tag{12.4}
$$

where y_j is the estimated log odds ratio in trial j with variance σ_{ej}^2 (which is assumed known). Under the assumption of normality, a confidence interval may be calculated for the average treatment effect θ.

For individual-level data, the conventional fixed effects model for p studies can be written as

$$
\left.
\begin{aligned}
\text{logit}(\pi_{ij}) &= \beta_0 + \sum_{k=1}^{p-1} \beta_k \delta_{ijk} + (\theta + v_j)x_{ij}, \\
v_j &\sim N(0, \tau^2), \\
y_{ij} &\sim \text{Binomial}(\pi_{ij}, 1),
\end{aligned}
\right\}
\tag{12.5}
$$

where y_{ij} is the binary response, π_{ij} is the probability of a positive response for the ith subject in the jth study, the δ_{ijk} are dummy variables for study membership, and x_{ij} is a dummy variable for treatment/control.

A particularly troublesome issue in all meta-analyses is that of publication bias, whereby certain kinds of studies tend not to get published. To allow for this, it is common to assign to each study a weight as a function of the selection probability for that study. Such models require assumptions on the specific form taken by the selection probabilities, and may involve rather arbitrary decisions for which robustness is lacking (Hedges and Vevea, 1996). Copas (1999) has recommended a sensitivity approach to the problem of publication bias, as an alternative to explicit estimation of corrected estimates. The proposed method involves examination of the extent to which the estimation of θ depends on parameters describing the selection probabilities. This procedure yields a range of plausible estimates of θ rather than a single corrected estimate, and sensitivity analyses using this procedure would seem to be useful.

12.3   SURVIVAL DATA MODELLING

This class of models, also known as event duration models, have as the response variable the length of time between 'events'. Such events may be, for example, birth and death, or the beginning and end of a period of employment, with corresponding times being length of life or duration of employment. There is a considerable theoretical and applied literature, especially in the field of biostatistics, and a useful summary is given by Clayton (1988).

The multilevel structure of such models arises in two general ways. The first is where we have repeated durations within individuals. Thus individuals may have repeated spells of various kinds of employment, of which unemployment is one, or women may have repeated spells of pregnancy. In this case, we have a two-level model with individuals at level 2, often referred to as a renewal process. We can include explanatory dummy variables to distinguish different kinds or 'states' of employment or pregnancy, such as the sequence number. The second kind of model is where we have a single duration for each individual, but the individuals are grouped into level-2 units. In the case of employment duration, the level-2 units would be firms or employers. If we had repeated measures on individuals within firms then this would give rise to a three-level structure

A characteristic of duration data is that for some observations we may not know the exact duration but only that it occurred within a certain interval. This is known as interval censored data: if less than a known value, left censored data; if greater than a known value, right censored data. For example, if we know at the time of a study that someone began her pregnancy before a certain date then the information available is only that the duration is longer than a known value. Such data are known as right censored. In another case, we may know that someone entered and then left employment between two measurement occasions, in which case we know only that the duration lies in a known interval.

There are a variety of models for duration times, and we here mention only three of the most common. We shall merely sketch the model without going into details of estimation. A full description of estimation procedures is given by Goldstein (1995).

Perhaps the most commonly used is the proportional hazards model, also known as a semiparametric proportional hazards model. Consider the two-level proportional hazards model for the jkth level-1 unit:

$$h(t_{jk};X_{jk}) = \lambda(t_{jk}) \exp(X_{jk}\beta_k), \qquad (12.6)$$

where X_{jk} is the row vector of explanatory variables for the level-1 unit and some or all of the β_k are random at level 2.

We suppose that the times at which a level-1 unit comes to the end of its duration period or 'fails' are ordered, and at each of these we consider the total 'risk set'. At failure time t_{jk}, the risk set consists of all the level-1 units that have been censored or for which a failure has not occurred immediately preceding time t_{jk}.

Another model in common use is the accelerated life model, where the distribution function for duration is commonly assumed to be of the form

$$f(t;X,\beta) = f_0(te^{X\beta})e^{X\beta},$$

where f_0 is a baseline function (Cox and Oakes, 1984). For a two-level model, this can be written as

$$l_{ij} = \log t_{ij} = X_{ij}\beta_j + e_{ij}, \qquad (12.7)$$

which is in the standard form for a two-level model. We shall assume normality for the random coefficients at level 2 (and higher levels), but at level 1 we may have other distributional forms for the e_{ij}. The level-1 distributional form is important where there are censored observations.

The third model, often used in demographic studies (Steele *et al.*, 1996), is the piecewise duration model. We suppose that the total time interval is divided into short intervals during which the probability of failure, given survival up to that point, is effectively constant. Denote these intervals by t $(1, 2 \ldots, T)$. We define the hazard at time t as the probability that, given survival up to the end of time interval $t - 1$, failure occurs in the next interval. At the start of each interval, we have a 'risk set' n_t consisting of the survivors, and, during the interval, r_t fail. If censoring occurs during interval t then this observation is removed from that interval (and subsequent ones) and does not form part of the risk set. A simple, single-level, model can be written as

$$\pi_{i(t)} = f[\alpha_t z_{it}, (\beta X)_{it}], \tag{12.8}$$

where $z_t = \{z_{it}\}$ is a dummy variable for the tth interval and α_t is a 'blocking factor' defining the underlying hazard at time t. The second term is a function of covariates. A common formulation would be the logit model, and a simple such model, in which the first blocking factor has been absorbed into the intercept term could be written as

$$\text{logit}(\pi_{i(t)}) = \beta_0 + \alpha_t z_{it} + \beta_1 x_{1i}, \qquad (z_2, z_3, \ldots, z_T). \tag{12.9}$$

Since the covariate varies across individuals, in general the data matrix will consist of one record for each individual within each interval, with a $(0,1)$ response indicating survival or failure. The model can be fitted using standard procedures, assuming a binomial error distribution.

As it stands (12.9) involves the fitting of $T - 1$ blocking factors. However, this can be avoided, (Goldstein, 1995, Chapter 9) by fitting a low-order polynomial to the sequentially numbered time indicator, $Z^* = 1, 2, \ldots, T$, so that (12.9) becomes

$$\text{logit}(\pi_{i(t)}) = \beta_0 + \sum_{h=1}^{p} \alpha_h^* (z_{it}^*)^h + \beta_1 x_{1i}, \tag{12.10}$$

where p is typically 3 or 4.

The logit function can be replaced by, for example, the complementary log–log function, which gives a proportional hazards model, or, say, the probit function. We note that we can incorporate time-varying covariates such as age. A 'competing risks' model with several different kinds of survival can be constructed by extending the response to become a multinomial vector representing the various risks.

Consider the two-level extension where we suppose that level 1 is individual (pregnancy length) and level 2 is community. A simple generalisation is

$$\text{logit}(\pi_{ij(t)}) = \beta_0 + \sum_{h=1}^{p} \alpha_h^* (z_{it}^*)^h + \beta_1 x_{1ij} + u_j, \tag{12.11}$$

where u_j is the 'effect' for the jth community, and is typically assumed to be distributed normally with zero mean and variance σ_u^2. We can elaborate this using random coefficients, resulting in a heterogeneous variance structure, further levels of nesting etc. This is just a two-level binary response model and can be fitted using, for example, quasi-likelihood or Markov-chain Monte Carlo (MCMC) methods (for details about using these in MLwiN, see Rasbash *et al.*, 1999a, b). The data structure has two levels, so that individuals will be grouped (sorted) within communities, but within each community the record order is again immaterial. For the competing risks model we use the multinomial two-level formulation (Goldstein, 1995). The setting up and fitting of such a model in MLwiN is described in Yang *et al.* (1999).

12.4 CONTEXT AND COMPOSITION

Multilevel modelling has been an important advance in health service and public health research since it has enabled a focus on both microlevel and macrolevel relationships simultaneously, as well as the relationships between them (Groenewegen, 1997). The questions facing researchers concern the degree to which observed differences at the macrolevel – typically hospitals or areas – reflect genuine contextual differences between those areas or whether they do little more than reflect the composition of those areas in terms of the microlevel (typically the individual). For example, Jones (1997) questions whether the relationship between voting behaviour and place is *contextual* – meaning that '... something about the social and economic milieux of [an] area ... produces a distinctive political culture' – or whether it merely reflects the *composition* of an area, with particular relevance to social class composition.

Duncan *et al.* (1998), discussing institutional performance (see Chapter 9 for further discussion of this subject), suggest that the average performance of a clinic can be seen to comprise three elements:

average clinic performance	=	composition of the clinic	+	contextual clinic difference	+	composition/ contextual interaction.

The composition of the clinic in this example refers to the make-up of the clinic in terms of the net characteristics of the people who attend the clinic; the contextual differences are the additional effect that a clinic has once its composition has been taken into account, and the interaction then reflects differential performance across patient groups. At a microlevel, the composition of the clinic could mean no more than taking individual patient characteristics into account; the interaction with the context then reflects a microlevel variable that is random across the macrolevel (clinics). However, this section considers compositional variables at both the micro- and macro-levels; that is, it considers the individual patient in relation to the overall composition of the practice.

Consider a hypothetical example in which the objective is to determine what effect different hospitals have on a patient outcome – for example, a rating of health following surgery. For every patient in every hospital data are collected as to their age, sex and whether or not the patient is receiving private medical care. Do these data then refer to patients or hospitals? Since they were collected for every patient, they must refer to the patient; however, in addition, they may provide hospital-level information. If the health system under study has two types of hospital – private and public – then this is a hospital-level variable. Moreover, if all patients receiving private medical care do so at private hospitals, and all other patients are treated at public hospitals, then there is no information at patient level (since every patient within each hospital will have the same classification). However, it may be that some privately funded patients receive their care in public hospitals; in this case, a comparison of interest may be between the outcomes of privately funded patients who are treated in public hospitals as opposed to those who are treated in private hospitals. Alternatively, the health system may have three types of hospital – private, public and mixed – the composition of the hospital can be seen separately from the individual-level variable from which it is derived.

In a similar manner, it may be important to draw comparisons between single-sex hospitals and mixed-sex hospitals (or hospital wards for a particular diagnosis), so sex may be considered a descriptor of hospital composition as well of the individual patients. This example can be developed further by considering the patient's age. This is not necessarily a question of comparing categories of hospitals – such as those providing paediatric or geriatric care – but may involve a more complex relationship between individual and hospital composition, such as the average age of the patients treated in that hospital. In this manner, the influence of the age of each patient on the outcome can be separated from the way in which it is influenced by the operational context of the hospital. Do older patients fare better in a hospital that predominately treats older patients, or one that generally treats younger patients? In a similar manner it is possible to consider the proportion of privately funded patients within a hospital rather than categorising all hospitals treating both types of patients as being mixed. The same is true for the patient's sex; in general, any microlevel variable – continuous or categorical – can also be considered at the macrolevel. The mean is a common way of summarising the data, but, depending on the particular research question, the minimum, maximum or another measure may be a more appropriate description of the composition of the higher-level units.

Duncan *et al.* (1998) give illustrations of a variety of ways in which individual and compositional variables may interact in cross-level relationships; these have been adapted in Figure 12.1. The two lines can be thought of as representing, by way of example, the predicted response for patients of different ages – say 50 years (broken lines) and 80 years (solid lines). With the vertical axis indicating the level of the response, the horizontal axis reflects the mean age of patients in the hospital. Figure 12.1(a) therefore illustrates a situation in which the 50-year-old patients are generally healthier than the 80-year-olds, and this difference is constant no matter what the composition of the hospital. Any differences between hospitals are therefore contextual rather than reflecting differences in composition. This is not to say that there is no difference between hospitals in terms of their composition, merely that their composition has no additional bearing on the patient outcomes. It may not be that the same average difference will be seen at all hospitals; it is possible that the age effect varies randomly across hospitals but there is no relationship between these random age effects and the hospital composition. Figure 12.1(b) suggests a scenario in which there is little difference between the health rating of the two age groups, but the ratings tend to be higher in hospitals in which the average age is high. So the health of the individual is determined not by the individual patient's age, but by the average age of patients treated in the hospital – this situation is the converse of Figure 12.1(a) in that it is the macrolevel compositional variable that is important rather than the microlevel patient characteristic. Figure 12.1(c) and (d) reflect a combination of these two situations in which both individual and compositional factors have an important impact on the response. The health of 50-year-olds is generally better than that of 80-year-olds in a hospital of the same composition, and this difference is independent of the hospital composition. Figure 12.1(c) presents a scenario in which the average health of all patients is improved in a hospital with a high mean age, whilst in Figure 12.1(d), the health of patients at hospitals with a young mean age is improved relative to patients of the same age who are treated in hospitals with a high mean age. The remaining four graphs illustrate some possible interactions between individual characteristics and hospital composition. In Figure 12.1(e), there is little difference between the health of those patients treated at hospitals with a high mean age, irrespective of the age of the individual patients. There are, however, substantial differences in hospitals with a low mean age, with younger patients faring very much better than the older patients. In Figure 12.1(f), the situation is the same in hospitals with a low mean age, but in hospitals with a high mean age it is the 80-year-olds whose health is better than the 50-year-olds. There are no age differences in the health of patients with composition in the middle of the age range. Figures 12.1(g) and (h) present more complex interactions; in both situations, the health of the 50-year-old patients is better than that of the 80-year-olds in hospitals with either a low or a high average age. However, in Figure 12.1(g) these differences disappear when patients are treated at hospitals with composition in the middle of the age range, with the average health of the older patients being better and that of the younger patients being worse; in figure 12.1(h), on the other hand, the differences between the two age groups are accentuated in these hospitals.

A common situation that gives rise to compositional effects is the differences between institutions in the recording of information. Leyland and Boddy (1998) considered the differences between Scottish hospitals in 30-day mortality rates following acute myocardial infarction (AMI). One patient characteristic strongly associated with death following AMI is the recording of a secondary diagnosis of other (non-ischaemic) heart disease (odds ratio = 1.77; 95% confidence interval (CI) 1.56–2.00). However, the large variation between hospitals

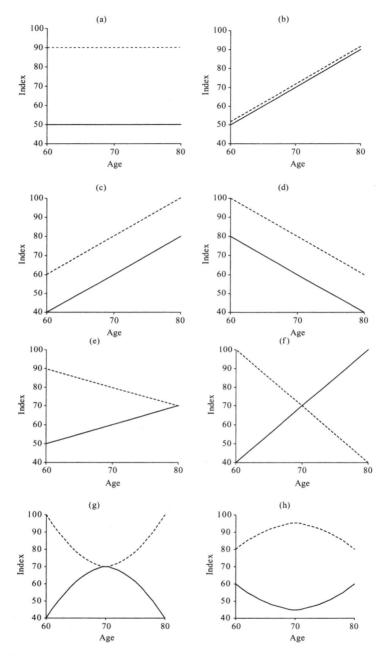

Figure 12.1 Illustration of cross-level interactions between individual and compositional variables. Reprinted from *Social Science and Medicine* **46**. Duncan, C., Jones, K. and Moon, G. Context, composition and heterogeneity: using multilevel models in health research. pp. 97–117. Copyright (1998), with permission from Elsevier Science.

in the proportion of patients for whom such a secondary diagnosis was recorded raises doubts as to whether this was really reflecting the patients' condition or the hospital's recording practice. Among the 31 hospitals that saw more than 50 AMI patients in the year of study, the recording of other heart disease ranged from 8% to 33%. If some hospitals have a lower threshold for a patient to be classified as having other heart disease (i.e. they are classifying a high proportion of patients in such a manner) then it is likely that the mortality rate among such patients will be lower in that hospital than in hospitals that have a higher threshold (those with a lower proportion of patients with other heart disease). However, if the classification of patients is still related to the severity of their condition then it is also likely that the mortality among patients who do not have a secondary diagnosis of other heart disease will also be lower among hospitals that have higher rates of recording of the condition. It is therefore possible to fit a model that indicates at the patient level the odds of mortality associated with the presence of other heart disease (1.74; 95% CI 1.51–1.98) and also the relationship between mortality and the composition of the hospital such that a 10% increase in the proportion of patients with other heart disease is associated with an odds ratio of 0.85 (0.72–1.01) among patients with other heart disease and 0.76 (0.68–0.86) among patients not classified as having other heart disease.

Table 12.1 shows the combined effect of other heart disease as both a patient characteristic and as a compositional variable. Three levels of the percentage of patients with other heart disease recorded are used: 8.8%, 17.0% and 32.7%, which correspond to the 5th, 50th and 95th percentiles for patients (that is, 5% of all patients are in hospitals in which no more than 8.8% have a recorded secondary diagnosis of other heart disease, etc.). The reference category is patients without a diagnosis of other heart disease in a hospital that records 17.0% of cases as having this diagnosis. The odds of mortality associated with other heart disease are 1.75 in such a hospital; this figure decreases to 1.58 in hospitals with the lower level of recording, and increases to 2.09 in hospitals recording 32.7% of patients as having other heart disease. This therefore corresponds to a situation somewhat similar to Figure 12.1(e), but with the lines diverging rather than converging as the compositional variable increases. Patients with other heart disease are always at greater risk than patients without this recorded (if the vertical axis is the odds of mortality, the broken line

Table 12.1 Odds of mortality (95% confidence intervals in parenthese) associated with hospital composition (percentage of patients with other heart disease) and the recording of other heart disease for individual patients.

Percentile	Percentage of patients with other heart disease	Patients without other heart disease	Patients with other heart disease
5th	8.8%	1.25 (1.14–1.38)	1.98 (1.64–2.45)
50th	17.0%	1.00	1.75 (1.51–1.98)
95th	32.7%	0.65 (0.54–0.79)	1.36 (1.06–1.76)

corresponds to patients with this diagnosis). In both groups of patients, the odds of mortality decrease as the compositional variable increases (so both lines slope down from left to right), and the increased risk becomes more pronounced as the level of recording of the diagnosis in a hospital increases (so the distance between the two lines increases).

12.5   RECENT DEVELOPMENTS IN MULTILEVEL MODELLING

The topic of measurement errors is a complex one, and there appears to be little attempt to make adjustments when models include such errors in either the response or predictor variables. It is known, however, that, in the presence of such measurement error, inferences may be biased. We refer the reader to a detailed discussion, with an example taken from education, given by Woodhouse et al. (1996).

One area that has been receiving some attention recently is multilevel covariance structure analysis or structural equation modelling. Hox (1995) describes this general approach as one that encompasses both path models and factor models; the former structural model is used to describe predictive relationships between observed variables and latent factors, whilst the latter factor model describes the construction of the latent factors from the observed variables. For introductory texts on this subject the reader is referred to Muthén (1994) and McDonald (1994). Software capable of fitting multilevel structural equation models includes Mplus (Muthén and Muthén, 1998) and STREAMS (Gustafsson and Stahl, 1997).

A further common issue in statistics concerns the analysis of data sets where some of the data are missing, and this concern extends to multilevel data sets. The fact that a balanced design is not a prerequisite of a multilevel model means that subjects with some outcomes missing may still be included in an analysis (see Chapter 5 on multivariate regression analysis). When the explanatory variables are incomplete, there are typically two options as discussed in Goldstein (1995). If the data can be viewed as being missing completely at random (MCAR) or missing at random (MAR) conditional on other explanatory variables but independently of the outcome (Rubin, 1976) then a two-stage approach may be adopted. At the first stage a multivariate model is used to obtain predictions for all missing data values. The second stage resorts to multiple imputation (Rubin, 1987); a number of complete data sets are formed by repeatedly sampling from the predicted distributions of the missing values. The data sets are then analysed in turn, and the estimates obtained are based on data with the correct distributional properties. If the data are not missing at random – that is, the fact that the data are missing is related to the response and is in itself informative (see, e.g. Best et al., 1996) – it is common practice to model the missingness mechanism and then proceed as if the data were missing at random.

CHAPTER 13

Software for Multilevel Analysis

Jan de Leeuw
UCLA Department of Statistics, Los Angeles, CA, USA

Ita G.G. Kreft
Health and Human Services, CSLA, Los Angeles, CA, USA

13.1 INTRODUCTION

In this chapter we review some of the more important software programs and packages that are designed for, or can be used for, multilevel analysis. These programs differ in many respects. Some are parts of major statistics packages such as SAS or BMDP. Others are written in the macro language of a major package. And some are stand-alone special-purpose programs that can do nothing but multilevel analysis. We have been involved in a number of these comparisons before. The first (Kreft *et al.*, 1990), comparing HLM, ML3, VARCL, BMDP5-V and GENMOD, was published in Kreft *et al.* (1994). The second comparison (van der Leeden *et al.*, 1991), comparing HLM, ML3 and BMDP5-V on repeated measures data, was published in van der Leeden *et al.* (1996). We give both the reference to the internal report version and to the published version, because the unpublished version usually has much more material. Giving both references also shows the unfortunate time interval between the two, which is especially annoying in the case of software reviews. The reviews were summarised briefly in our book (Kreft and de Leeuw, 1998, Section 1.6).

Since our last publication on the subject, there have been many major changes. The program GENMOD, which was never easy to obtain, has more or less completely disappeared. VARCL, which was one of the leading contenders in the early 1990s, is no longer actively supported or developed, which means that it has rapidly lost ground. BMDP, as a company, went out of business, which had serious consequences for its software products. Programs such as HLM, written originally for DOS, were upgraded for Windows. ML3

Multilevel Modelling of Health Statistics Edited by A.H. Leyland and H. Goldstein
©2001 John Wiley & Sons, Ltd

transformed to MLn, which transformed to MLwiN. And so on. Obviously, it is time for an update. In this update, we shall change our focus somewhat. We shall indicate what the programs can do, where you can get them, for how much money, on what systems they run, and how easy it is to use them. We do not emphasise computation speed, because this is hard to define, and not very relevant in most applications anyway. Computing time is usually infinitely small compared with the time needed to collect and clean the data.

13.1.1 Programs, macros and packages

Throughout this chapter, we shall use a number of classifications of the software we review. Of course the boundaries between the categories are somewhat fuzzy.

Research versus production Some software is written as a research tool, to analyse the examples in a research paper, or to analyse the data sets in a research project. Other software is written as a commercial product, or at least it is clearly intended for general use. Many programs start out in the research phase, and are subsequently promoted to the production phase. Many programs are still somewhere in between.

Stand-alone versus module There are multilevel programs that stand on their own, i.e. they are executables and they do not require other software (except the operating system, of course) to be present. Other programs are modules of an existing package. Usually they require the rest of the package to be present as well.

Program versus macro For programs written on top of an existing package, there are still two possibilities. Either the program is a module, existing in object code within the package, or the program is a macro, written in the scripting or extension language of the package. Macros are handled by an interpreter that is part of the package.

Another distinction that we make is between minor and major specialised programs. This is somewhat subjective, but the general idea is that a research program, usually with a rather primitive interface, is a minor specialised program. As soon as the authors start thinking about user-friendliness, making a nice graphical user interface, adding options that people have asked for, and going commercial, the minor program will become a major one. It will become bigger, have more possibilities, look better, and cost more.

13.1.2 Omissions

There are some programs that can be used to perform multilevel analysis, but that we do not discuss in detail. We just mention these programs briefly here, and we give the URL in case readers want to know more. (The http:// part of the URLs have been omitted throughout.)

TERRACE   This is multilevel research software, written by James Hilden-Minton for his PhD thesis. It is an add-on to the XLISP-STAT package, and it can be found (with manual) at
www.stat.ucla.edu/consult/paid/nels/papers

NLME   The NLME software comprises a set of S-plus functions, methods and classes for the analysis of both linear and nonlinear mixed-effects models. It extends the linear and nonlinear modelling facilities available in release 3 of S-plus. It was written by José C. Pinheiro and Douglas M. Bates. It is available for Unix and Windows platforms from
franz.stat.wisc.edu/pub/NLME

BUGS   This is a piece of computer software that permits the analysis of complex statistical models using Markov-chain Monte Carlo methods. The emphasis is on the Monte Carlo method, and a great variety of multilevel models can be analysed as well (see, e.g. Chapter 9 on institutional perform-ance). For further details, see
www.mrc-bsu.cam.ac.uk/bugs/welcome.shtml

Oswald   This was developed by the Statistics Group at the University of Lancaster, and is a suite of S-plus functions for analysing longitudinal data. It includes mixed effects models, and many other possible options.
www.maths.lancs.ac.uk/Software/Oswald/

13.2   MAJOR SPECIALISED PROGRAMS

There are two major specialised programs for performing multilevel analysis – one from the UK and one from the USA.

13.2.1   MLwiN

The ML series of programs has a complicated history. The series was erected on top of the NANOSTAT program by Michael Healy. NANOSTAT is a general-purpose statistics program. The multilevel extensions started with ML2 in 1988, ML3 was introduced in 1990, and the final DOS program in the series was MLn, published in 1995. In 1998, they were all superseded by MLwiN. One can think of MLwiN as a separate program, but also as a graphical user interface on top of MLn. Throughout the project, most of the programming was done by Jon Rasbash, but clearly the program is the result of a team effort.

MLwiN contains the NANOSTAT package, so it can do a fair amount of data manipulation and general purpose statistics. This is all meant to assist in the multilevel analysis, and thus we still think of MLwiN as a specialised program.

In our report (Kreft *et al.*, 1990) we looked at ML2, and in the published version (Kreft *et al.*, 1994) at ML3. In a subsequent comparison, the internal report version (van der Leeden *et al.*, 1991) looked at ML3, and the published version (van der Leeden *et al.*, 1996) at MLn.

We shall only discuss MLwiN in this chapter, since it includes all previous programs as special cases, and since most people seem to feel that putting a Windows interface on top of a DOS program is a step ahead.

Availability　The Multilevel Models Project has three mirror homepages at
`www.ioe.ac.uk/multilevel/`
`www.medent.umontreal.ca/multilevel/`
`www.edfac.unimelb.edu.au/multilevel/`
The websites have information about the software, but also about the project and its activities. The MLwiN program has its own homepage at
`www.ioe.ac.uk/mlwin/`
The price of a single copy given on the MLwiN website is £500 ($900), although there is a 40% educational discount.

Documentation　There is a large amount of documentation available, most of it is can be obtained from the multilevel project. Again, the various versions of the program lead to a somewhat bewildering variety of documents. Basically, there are three types of documents available for most versions of the program. The first document is the user's guide. For MLwiN, this is Goldstein *et al.* (1998). There is another type of user's guide, which is more introductory. For MLn, this is Woodhouse (1995). The second is the advanced manual, which discusses macros to fit more complicated models. This is Yang *et al.* (1998). Thirdly, there is the command reference. Since the command language for MLwiN is actually MLn, this is Rasbash and Woodhouse (1995).

In a sense, the most important document in the Multilevel Models Project is the book by Harvey Goldstein. Not surprisingly, this also exists in two very different editions. It developed with the project. The first edition (1987) covers basic multilevel analysis, with emphasis on applications in education. The second edition (1995) is much more statistical, and covers a large variety of extensions of the basic model and a much broader range of applications.

Model　The basic model, in the case of two levels and p predictors, is

$$y_{ij} = \beta_{0ij}x_{0ij} + \beta_{1ij}x_{1ij} + \ldots + \beta_{pij}x_{pij}, \tag{13.1}$$

where

$$\beta_{sij} = \beta_s + u_{sj} + e_{sij}. \tag{13.2}$$

It is easy to see how this generalises to more than two levels. Observe that usually we have $x_{0ij} = 1$ for all i, j, i.e. the zero term in the regression corresponds to the intercept. Also observe that each regression coefficient has a fixed part and a random part, and the random part has random components for both levels.

Interface　As indicated above, MLwiN provides (or maybe we should say 'is') a Windows interface on top of MLn. It is possible to use MLwiN very much like MLn, because there is a command window in which MLn commands can be entered. This window also tracks the command history. In fact, menu commands are translated to MLn commands in the command window. There are many MLn and NANOSTAT commands that *cannot* be entered from the menus.[1]

An interesting feature of the program is the *equations window*. This is a specialised equations editor, in which the model can be defined in equation form by clicking and pasting. Obviously, there is no way to do this in DOS, so this defines a clear distinction between MLn and MLwiN. On the other hand, it is unclear if this is actually superior to written output that users themselves have to translate into a formula. It seems that if users know what they are doing, they can make the translation very easily themselves. If they don't know what they are doing, then they should not be using the program in the first place!

Algorithm MLwiN uses the IGLS or the RIGLS algorithms first described respectively by Goldstein (1986) and Goldstein (1989b). The algorithms are block-relation algorithms. There are two blocks of parameters: the fixed regression coefficients and the variance/covariance components. The algorithm fixes the variance components at some initial value and maximises the likelihood over the fixed coefficients. This is just a generalised least squares problem. Then it fixes the coefficients at their current values and maximises the likelihood over the variance components, by solving another, more complicated, generalised least squares problem. The two optimisations are alternated until convergence. A concise description appears in Goldstein and Rasbash (1992).

It is not entirely clear from the documentation what happens in boundary cases when dispersion matrices become singular or even indefinite.

Extensions By 'extensions' we mean various options and additions that do not really belong to the core of the program, but that the authors have added because of user demand, competitive pressure or their research program. Some extensions are obviously more useful for the general public than others, but most of them are at least interesting enough to be mentioned.

In MLwiN, there are two 'levels' of extensions. The first are features that are part of the program core. They can be handled by using the menus, or the MLn language, but they are features that will not often be needed. The second level consists of true extensions, written in the macro language provided by the package, and these are add-ons that the user may or may not load.

The first class of extension is discussed in the user's guide (Goldstein *et al.*, 1998). This already covers a substantial number of procedures. Hierarchical generalised linear models with binomial or Poisson outcomes can be fitted. Markov-chain Monte Carlo methods are available to optimise complicated likelihoods or compute complicated posterior distributions. Parametric

---

[1] Editor's note: The current release of MLwiN, v1.10.000 at the time of going to press, has many more MLn/NANOSTAT comands included in the menus.

bootstrap methods are used for bias correction and for standard error computation.

These extensions may be exciting to some, but they do take MLwiN several steps in the direction of research software. Users have to take very many things for granted, and have to hope that the default values of the many parameters and tuning constants work in their case. It seems a bit too demanding to ask casual users to choose between the MQL and PQL methods for quasi-likelihood estimation (see Chapter 3 for a discussion of these estimation procedures), or to choose an appropriate burn-in period for their MCMC procedure. It is true that with these extensions, models can be fitted that could not be fitted by older versions of MLn. But, for most people, it becomes impossible to understand what is actually going on inside the program, and to explain why certain choices and not others were made.

The macro-based extensions of MLwiN allow for even more flexibility. Macros are available for fitting multicategory models (see Chapter 8), survival models (see Chapter 12), time-series models (see Chapter 2) and nonlinear models (see Chapters 3 and 4). While these extensions are undoubtedly useful, again the reservations in the previous paragraph apply.

For the average MLwiN user, the instructions in the manual are voodoo. There are references to the statistical literature, of course, but these references are in many cases too technical to be of much use.

This, of course, is a well-known dilemma. If applied researchers, with often quite limited technical expertise, want to fit very complicated models with very complicated algorithms then the documentation and the implementation should handle this very carefully. Both in Goldstein (1995) and in the various manuals, one often gets the idea that instead of carefully guiding users through complicated territory, they are invited to climb on the roller coaster, close their eyes, and enjoy the ride.

13.2.2   HLM

HLM's history is similar to MLwiN's. There was first an HLM for two-level models, then one for three-level models, then one that could also do generalised linear regression, and finally a Windows version. This last version, version 4 of HLM/2L and HLM/3L, is the one we review here. Of course HLM/2L does two-level analysis and HLM/3L three-level analysis. The programming was done by Richard Congdon; the HLM team also includes Stephen Raudenbush and Tony Bryk.

Availability   The software is available from Scientific Software International in Chicago, where HLM has its homepage
www.ssicentral.com/hlm/mainhlm.htm
The price is $400 for the DOS version and $430 for the Windows version.

Documentation   The HLM documentation consists of a user's manual (Bryk *et al.*, 1996) and a book by Bryk and Raudenbush (1991). In a sense, the book is independent of the software, but since you get the book when you buy the software, and since the HLM program is used throughout the book, the two are really intimately related (even closer than MLwiN and Goldstein (1995)). The choice of name HLM can lead to unfortunate confusion of the model, the technique and the package (de Leeuw and Kreft, 1995).

The Windows interface to HLM is only documented in the on-line help, it seems.

Model   The basic HLM model, for two-level data, is

$$y_{ij} = \beta_{0j} x_{0ij} + \beta_{1j} x_{1ij} + \ldots + \beta_{pj} x_{pij} + e_{ij}, \tag{13.3}$$

where

$$\beta_{sj} = \gamma_{s0} z_{j0} + \gamma_{s1} z_{j1} + \ldots + \gamma_{sh} z_{jh} + u_{sj}. \tag{13.4}$$

To make the comparison with MLwiN somewhat easier, we rewrite this as

$$y_{ij} = \gamma_{00} x_{0ij} z_{j0} + \ldots + \gamma_{ph} x_{pij} z_{jh} + u_{0j} x_{0ij} + \ldots + u_{pj} x_{pij} + e_{ij}. \tag{13.5}$$

There are two obvious differences between the programs.

1. In MLwiN, we can have random coefficients β_{sij} on the first level. This is not possible in HLM.
2. In HLM, the emphasis is on the cross-level interactions in the fixed part, which are products of a first-level and a second-level predictor. It is possible to have such variables in MLwiN, but they are not as central.

Interface   Similar to what we saw in MLn, the newer versions of HLM now have a Windows interface. In addition, they have the interactive (question-and-answer) and batchfile interfaces from the older versions. In Bryk *et al.* (1996), the authors remark that most PC users will prefer the Windows interface – but it is pretty clear from the rest of the book that they themselves do not.

The Windows version of HLM also comes with an equation editor, similar to the one in MLwiN.

Algorithm   By default, HLM/2L uses REML estimation, while HLM/3L uses FIML. Nevertheless, both programs can actually do both forms of estimation. Older versions of HLM relied on the EM algorithm, which can sometimes be hopelessly slow. It is now possible to speed up convergence by switching to Fisher scoring.

Extensions   There are a number of interesting extension in HLM. First, there is the V-known option (where the variance components are supposed to be known), which can be used in meta-analysis. Secondly, there are hierarchical generalised linear models, fitted by using penalised quasi-likelihood or generalised estimation equations. In particular, Poisson, Bernoulli and binomial models can be fitted. Thirdly, a form of plausible value analysis using multiple imputation is available for two-level models. Although the scope of HLM is obviously much smaller than that of MLwiN, this restriction of generality makes the program easier to use. Generally, the documentation is more user-oriented, the number of choices the user can make is more limited, and often the authors of the program have already made many of the choices.

13.3   MINOR SPECIALISED PROGRAMS

13.3.1   VARCL

The program VARCL started out in the mid-1980s as one of the major contenders. It was Longford's research software, used in the path-breaking paper by Aitkin and Longford (1986) and in his book (Longford, 1993). Since 1990, Longford has moved to using S-plus for research software development, and nothing has happened with VARCL. It is still a DOS program.

Availability   The program is sold by ProGAMMA in Groningen, The Netherlands, but it is in a very remote section of their catalogue:
www.gamma.rug.nl
The price is $375 ($250 educational).

Documentation   The distribution comes with a 100–page user manual (Longford, 1990), with an additional 100 pages of example runs. The manual follows the interactive interface closely, but 20 pages are used for explaining some of the technical background.

Model   There are two different versions of VARCL available. The first is VARCL3, which handles three-level models, the second is VARCL9, which handles random intercept models with up to nine levels. The model in VARCL3 is

$$y_{ijh} = \beta_{0jh}x_{0ijh} + \beta_{1jh}x_{1ijh} + \ldots + \beta_{pjh}x_{pijh} + e_{ijh}, \qquad (13.6)$$

where

$$\beta_{sjh} = \beta_s + u_{sh} + u_{sjh}. \qquad (13.7)$$

Again, we use the notation of MLwiN, and we assume that $x_{0ijh} = 1$ for all i, j, h.
   In VARCL9, the model is (for four levels, as an example)

$$y_{ijhg} = \alpha_{jhg} + \beta_1 x_{1ijhg} + \ldots + \beta_p x_{pijhg} + e_{ijhg}, \qquad (13.8)$$

where

$$\alpha_{jhg} = \alpha + u_g + u_{hg} + u_{jhg}. \qquad (13.9)$$

Here we have singled out the intercept more explicitly, by using α for the corresponding regression coefficient.
   We see that VARCL does not have the emphasis on cross-level interactions (or on two-level specification) that HLM has. It also does not have the facility for random coefficients at the first level, which does exist in MLwiN.

Interface   VARCL has a command-line interface. The programs asks a large number of questions, and it uses the information provided by the user to build up a setup file that describes the analysis. It first builds up information about the maximal model, which is the largest model (in terms of the number of variables) that can be fitted in a session. Then additional questions are used to exclude variables from the maximal model to define the model to be fitted. Constraints on the parameters can be defined, by setting them to fixed values. The model is shown, and, if the user likes it, the program computes the estimates. Then the interface asks if additional models will be fitted, and if so, which variables are to be removed and added, and which parameters have to be constrained or freed.

If you know from experience which questions the program will ask, and which answers you are going to give, then you can put the answers in a batch file and take standard input from that file.

Algorithm   The program uses the scoring method on the reduced form. It is based on the fact that within-group means, variances and covariances are jointly sufficient for the model parameters. Thus, if we compute the within-group statistics for the maximal model, we can forget about the original data. Since these within-group statistics can be computed in the input loop, the number of observations that VARCL can handle is infinitely large, as long as they come in a finite number of groups. This is different from MLwiN, which keeps the data in core for the whole session. It is similar to HLM, which uses the within-group statistics to compute separate regressions for each group (if the groups are big enough). Both VARCL and HLM use within-group statistics to compute initial estimates of the variance and covariance components.

In our experience, documented in our previous review papers, the scoring algorithm is both fast and reliable.

Extensions   VARCL was also the first multilevel program that could deal with hierarchical generalised linear models. Early on in the interactive questions and answers session, the user can choose the error model to be normal, binomial, Poisson or gamma. Again quasi-likelihood methods are used to fit these models, using the procedures designed by Longford (1988) for this purpose.

13.3.2   MIXFOO

The MIXFOO program is special – because it does not exist. It is the generic name for a whole series of multilevel programs. At the MIXFOO website, we find the following: 'The statistical research presented in this homepage is based on the collaborative effort of Donald Hedeker and Robert D. Gibbons of the University of Illinois at Chicago. The computer programs were written by Don Hedeker with interfaces written by Dave Patterson (Discerning Systems, Inc.).'

The names of all programs in the series start with MIX. There are DOS programs MIXOR, MIXREG, MIXGSUR, MIXNO and MIXPREG, while MIXOR and MIXREG also exist with Windows interfaces. The website also has a SAS-IML macro that fit a random intercept version of MIXREG, and SPSS-MATRIX macros for random intercept versions of both MIXOR and MIXREG. All this is freely available from the website.

We give here a brief indication of what these programs do. MIXREG fits the linear multilevel model, but it allows for various forms of autocorrelation between the first-level disturbances. MIXOR adds ordinal multicategory outcomes. MIXGSUR fits grouped-time survival data, MIXNO fits nominal multicategory data, and MIXPREG handles multilevel Poisson regression.

Availability   Software and manuals can be downloaded from
www.uic.edu/~hedeker/mix.html
All programs, macros and manuals are free. There are PowerMac and Sun/Solaris version of the software at
www.stat.ucla.edu/~deleeuw/mixfoo
It must be emphasised that this is quite unique – all other programs (except VARCL and MLA) only exist for DOS or Windows.

Documentation   The two core programs, MIXOR and MIXREG, are described in Hedeker and Gibbons (1996a) and Hedeker and Gibbons (1996b). These are published versions of the manuals; the manuals themselves are included in the software distribution. The theory is described in great detail in Hedeker (1989) and for MIXOR in the article by Hedeker and Gibbons (1994). MIXGSUR, MIXNO and MIXPREG also have a manuals on the website. The theory of MIXGSUR is described in the technical report by by Hedeker *et al.* (1996). More generally, there is a list of both theoretical and applied articles using MIXFOO at
www.uic.edu/~hedeker/works.html

Model   For MIXREG, the model is

$$y_{ij} = \alpha_0 w_{0ij} + \ldots + \alpha_p w_{pij} + \beta_{0j} x_{0ij} + \ldots + \beta_{qj} x_{qij} + e_{ij}, \qquad (13.10)$$

with

$$\beta_{sj} = \beta_s + u_{sj}. \qquad (13.11)$$

This is a straightforward two-level random coefficient model, very similar to what we have in VARCL. The unique aspect is that we do not assume that

$$\text{cov}(e_{ij}, e_{kj}) = \delta^{ik}\sigma^2, \qquad (13.12)$$

but we allow for a much more general parametric first-level error covariance structure. More specifically, e_{ij} the can have an autoregressive AR(1), moving-average MA(1), or autoregressive moving-average ARMA(1,1) covariance structure, a general stationary autocorrelation structure, and even a special non-stationary one.

For MIXOR, the model is the same, except that we do not observe the y_{ij}; rather, we observe a multicategory version generated by a threshold model. Thus there are unknown cut-off points (assumed to be the same for all variables). If y_{ij} is below the first cut-off, we observe a '1'; if it is between the first and second cut-offs, we observe a '2'; and so on. Thus the model is basically the same as before, but there are 'missing data' because we only know that y_{ij} is in a particular interval, but we don't know where it is in that interval. More precisely, the model is

$$\eta_{ij} = \alpha_0 w_{0ij} + \ldots + \alpha_p w_{pij} + \beta_{0j} x_{0ij} + \ldots + \beta_{qj} x_{qij} + e_{ij}, \tag{13.13}$$

where η_{ij} is the unobserved continuous response, and

$$y_{ij} = \begin{cases} 1 & \text{if } k_0 \le \eta_{ij} < k_1, \\ 2 & \text{if } k_1 \le \eta_{ij} < k_2, \\ r & \text{if } k_{r-1} \le \eta_{ij} < k_r, \end{cases} \tag{13.14}$$

where $k_0 = -\infty$ and $k_r = +\infty$.

Interface   All programs exist as batch versions, using command files. MIXOR and MIXREG also have interactive DOS versions, in which the command file is constructed from menu commands issued by the user. Finally, MIXREG and MIXOR are available as Windows programs.

Algorithm   MIXREG uses a combination of the EM and the scoring algorithm, in much the same way as, for instance, HLM. For MIXOR there are additional complications, because multidimensional integrals must be evaluated to compute the likelihood and its derivatives. MIXOR approximates these integrals by using Gauss–Hermite quadrature.

Extensions   If we consider MIXREG and MIXOR to be the basic components then MIXGSUR, MIXNO and MIXPREG are extensions. But of course drawing the line in this way is rather arbitrary. It is better to think of the whole set of programs as a modular alternative to programs such as HLM and MLwiN, which cover about the same amount of territory in a single program.

13.3.3   MLA

MLA is a batch program, running under DOS. It was written by Frank Busing, Rien van der Leeden, and Eric Meijer of the University of Leiden, The Netherlands. It differs from similar programs because it has extensive simulation possibilities built in (notably the bootstrap and the jackknife), and it has various ordinary least squares estimation methods as options.

Availability   Software and manual can be downloaded from
`www.fsw.leidenuniv.nl/www/w3_ment/medewerkers/BUSING/MLA.HTM`
The program is free.

Documentation   The software distribution contains a 70–page Postscript manual (Busing *et al.*, 1994). It is a bit wordy, because it tries to spell out all details, especially the technical ones. The ultimate example is the 20–page

Technical Appendix A, which gives in painstaking detail the derivations of the formulae for the likelihood function, its first and second derivatives, and their expectations. It illustrates the effect TEX has on the mind of an individual who has just escaped from the dungeons of WYSIWYG. The user's guide portion of the manual is quite clear, however.

The authors have informed us that the manual is out of date, because many options have been added to MLA in the meantime. The program can now make histograms (of bootstrap results, for instance) and scatterplots (of residuals), and there are different bootstrap-based confidence intervals. Also, permutation tests for testing intraclass correlation are available.

Model The model is the same as in (13.3) and (13.4).

Interface MLA requires the user to create a parameter file, and then the program is started from the DOS command line as

$$\text{mla} < \text{inputfile} > \text{outputfile}$$

Not much of an interface – but it does the job.

Algorithm MLA uses the BFGS algorithm, which is a general-purpose quasi-Newton optimisation algorithm, to maximise either FIML or REML. Alternatively, it can also use EM. In order to make sure that the level-2 dispersion matrix is positive-semidefinite, two different parameterisations are available that ensure this.

Extensions MLA is somewhat unremarkable as a multilevel program (although the parameterisations of the dispersion matrix and the use of BFGS are unique). However, it is remarkable because it supports a wide variety of simulation analyses. This tends to suggest that it is research software, but for practitioners the confidence information will also be useful.

Here is a list of the unique features taken from the website:

- different kinds of simulations (bootstrap, jackknife and permutation),
- different methods of bootstrap simulation (cases, parametric and error),
- different types of residual estimation (raw and shrunken),
- different cases resampling schemes (level 1, level 2 and both),
- balanced resampling schemes (balanced bootstrap),
- linking of residual levels,
- distribution plots (histograms) for parameters, standard errors and t-value.

13.4 MODULES IN A MAJOR PACKAGE

13.4.1 BMDP5-V

BMDP was the first major statistical package. It has always had an excellent reputation, because its approximately 40 programs were developed in close

cooperation with excellent statisticians. But BMDP has a stormy recent past. The company was bought by SPSS in 1996. It is difficult to find BMDP on the main SPSS website at http://www.spss.com. It turns out that many of the BMDP products have been declared 'dead' by SPSS, and the only product still sold seems to be BMDP Classic for DOS. See
www.spss.com/software/science/BMDP
Some additional research suggests that SPSS plans to integrate BMDP in its SYSTAT product. In corporate-speak: 'I would like to welcome BMDP customers to the SYSTAT family.' In Europe, the situation is different. One can buy BMDP from Statistical Solutions in Cork, Ireland, and their web page clearly indicates they consider it to be one of their major products.

Availability BMDP can be ordered from
www.statsol.ie/bmdp.html
The price is £1295 (£1100 academic; $895 and $695 in the Americas), but of course this is for the whole package (all 40+ routines, of which BMDP5-V is only a single one).

Documentation The theory behind BMDP5-V is explained in Jennrich and Schluchter (1986). Very useful information about both the methodology and the program is in Schluchter (1988). There is also a chapter on 5V in the BMDP user's manual (Dixon, 1992) and the BMDP user's digest (BMDP Statistical Software, 1992). And, last but not least, there is a detailed 'how-to-use' book with many examples (Dixon and Merdian, 1992) (with a 5.25 floppy in the back!).

Model BMDP5-V is described in the documentation as a program for unbalanced repeated measures models with structured covariance matrices. This means that the emphasis is not on general hierarchies, but specifically on repeated measures, i.e. on repetitions nested within subjects.
To maintain the same notation as we have used before, we rewrite the model as

$$y_{ij} = x_{1ij}\beta_1 + \ldots + x_{pij}\beta_p + e_{ij}, \tag{13.15}$$

where the index j now stands for subjects, and i stands for the repeated measures (nested within subjects). Thus it seems there are no random coefficients. BMDP5-V allows a large number of possible choices for the within-subject covariance matrices Σ_j which has elements

$$\sigma_{ii'j} = E(e_{ij}, e_{i'j}). \tag{13.16}$$

We briefly describe the available options, in matrix notation:

$$\Sigma_j = \sigma^2 \mathbf{I}, \tag{13.17}$$

$$\Sigma_j = \omega^2 \mathbf{e}\mathbf{e}^{\mathrm{T}} + \sigma^2 \mathbf{I}, \tag{13.18}$$

$$\Sigma_j = \mathbf{Z}_i \mathbf{Z}_i^{\mathrm{T}} + \sigma^2 \mathbf{I}, \tag{13.19}$$

$$\mathbf{\Sigma}_j = \mathbf{\Lambda\Lambda}^{\mathrm{T}} + \mathbf{\Psi},\tag{13.20}$$

$$\mathbf{\Sigma}_j = \theta_1\mathbf{G}_1 + \ldots + \theta_r\mathbf{G}_r,\tag{13.21}$$

$$\mathbf{\Sigma}_j = \mathbf{\Sigma}.\tag{13.22}$$

Moreover, there are two options that can best be defined in elementwise notation:

$$\sigma_{ii'j} = \sigma^2\rho^{|i-i'|},\tag{13.23}$$

$$\sigma_{ii'j} = \theta_r, \quad \text{where} \quad r = |i - i'| + 1.\tag{13.24}$$

Clearly, this covers many of our well-known friends from earlier sections.

Model (13.17) defines independent observations (within subjects; there is always between subject independence), while model (13.18) (sometimes described as compound symmetry) is in fact the random intercept model. Model (13.19) is the random-coefficient model, which allows for random slopes as well. Thus random coefficients, which seemed to be missing in the BMDP5-V specification, enter again through the back door. Model (13.20) says that within-subject covariance matrices have a factor analysis structure, while (13.21) describes a general linear covariance structure, and (13.22) allows for any matrix (except that it must be the same over subjects). The final two models allow for autoregressive structures, similar to those we have seen in MIXREG. BMDP5-V also allows you to define your very own structure, provided you write a FORTRAN program to compute the structure and its derivatives. Really enterprising users could fit a repeated measures model in which the within-subject covariance structures satisfy a LISREL model, for instance.

Interface BMDP is a very modular package. Each of the 44 programs can basically be used in a stand-alone way. This makes it quite different from SAS, which is much more integrated. Modularity produces a great deal of efficiency. On the other hand, BMDP is very, very DOS. It comes with BMDP/DYNAMIC, a somewhat optimistically named data and program managing module. BMDP/DYNAMIC has the familiar DOS-type pseudo-menus and pseudo-windows. It is fairly lightweight in terms of overhead.

Algorithm A variety of algorithms are available. If you choose the FIML criterion, you can use Fisher scoring, Newton–Raphson or generalised EM. In the case of REML, generalized EM and quasi-scoring are available.

Extensions Some of the options for the within-subject covariance matrix are rather esoteric, and could be considered extensions. Certainly options for which you have to write the code yourself qualify as such.

13.4.2 PROC MIXED

SAS has many modules (or PROCs) to fit linear models and generalised linear models. Until fairly recently, however, there was nothing that could compete with BMDP5-V. But now there is – and with a vengeance. PROC MIXED is the most general program of those we discuss, and it is also embedded in the totality of SAS (or whatever parts of SAS you happen to have installed).

Availability   You can obviously order the package from SAS Institute. Start at www.sas.com/service/techsup/faq/stat_proc/mixeproc.html Because SAS is such a huge and complicated product, it does not make much sense to give a price. Many users will simply add PROC MIXED to their PC or mainframe SAS setup. Many people probably already have it available without actually knowing this.

Documentation   The documentation for PROC MIXED is very extensive. It consists of a chapter in SAS (1992). This is written in the familiar format of the SAS manuals. It has almost 100 pages, set up like a user's guide. There is not much theory, and not many examples, but the instructions and options are explained in detail.

There is also an excellent and extensive book by Littell *et al.* (1996). It has chapters on randomised blocks, repeated measures, multilevel models, analysis of covariance, spatial models, heteroscedastic models and best linear unbiased prediction. There are also chapters on generalised linear mixed models and nonlinear mixed models, where the SAS-IML macros GLIMMIX and NLIN-MIX are explained. The book contains many examples, worked out in detail.

For people who have started out with a program such as HLM, which has a much simpler interface and a much smaller menu of models, the use of PROC MIXED is explained expertly by Singer (1998).

In terms of documentation, PROC MIXED is the clear winner in our comparison. In most other respects, it is not.

Model   The model in PROC MIXED is very similar to the model in BMDP5-V. It may even be true, historically, that BMDP5-V inspired PROC MIXED. The main difference, however, is that the PROC MIXED specification does not have levels or repeated measures. They start from the general mixed model, which is

$$y_i = \beta_1 x_{1i} + \ldots + \beta_r x_{ri} + u_1 z_{1i} + \ldots + u_s z_{si} + e_i. \tag{13.25}$$

The random effects u have a covariance matrix Σ and the errors e have a covariance matrix Δ. As in BMDP5-V, we can specify additional parametric structure in these covariance matrices. PROC MIXED allows for uncorrelated, compound symmetry, unstructured, autoregressive, and spatial (or a block-diagonal version of these – which is how the levels come in again). There is no factor-analytic or general linear covariance structure.

It is interesting that HLM and MLA fit true multilevel models, in the sense that they specify their regression model at multiple levels. MIXFOO and MLwiN have multiple levels, but only a single regression equation. The same is true for BMDP5-V, where the levels are defined by repeated measures. In PROC MIXED the levels have disappeared, and they have to be introduced by suitably arranging the input and parameter files. This is precisely the reason why a paper such as Singer (1998) is necessary for some groups of users.

Interface The interface to PROC MIXED is the familiar SAS interface – or more precisely any one of the familiar SAS interfaces. As in the case of BMDP, all these interfaces are rather desperate attempts to get away from the mainframe or DOS heritage. Most of the Windows interface is used to construct batch files with options and instructions, which are then submitted to the old familiar SAS engine in the background.

Algorithm PROC MIXED can use both REML and FIML estimation, and it maximises the likelihood by a combination of Fisher scoring and Newton–Raphson. The default is to use only Newton–Raphson, but optionally one can start with some Fisher scoring iterations.

Extensions The two major extensions of PROC MIXED are both written in the macro language SAS-IML, by Russ Wolfinger. They are discussed in detail in Littell *et al.* (1996, Chapters 11 and 12). Generalised linear mixed models can be fitted with the GLIMMIX macro. This incorporates binomial models with logit and probit links and Poisson (count) models with the log link. The macros uses a joint quasi-likelihood method. Nonlinear mixed models can be fitted with the NLINMIX macro. This macro actually has three options, depending on whether first-order, second-order or marginal approximations are used. This allows one to fit a really large range of models such as nonlinear growth curve models.

It is probably obvious that using such macros cannot possibly be very efficient. Using SAS to fit your multilevel models is already quite a stretch, because you have to carry along the enormous overhead of the SAS system. A class of students, each using SAS on a single CPU, will show you what this means. On top of that, the use of macros and the SAS-IML interpreter adds another layer of inefficiency. It's a little bit like using a big truck to pick up some groceries at your local supermarket.

There is a trade-off, of course. Many people have SAS installed and are quite familiar with its interface and macro language. If they need to fit a multilevel model once in a while, it probably does not make sense for them to learn a new language or to buy a new software package. If you are already driving around in a truck, you might as well pick up your groceries.

13.5 CONCLUSIONS

We have already commented throughout the text on some of the differences. There are multilevel, random coefficient and mixed linear model programs. There are package modules and stand-alone programs. There are research and production programs. One obvious characteristic they have in common is that they are almost all DOS, trying to become Windows.

We do not really want to get too deeply into the discussion of the relative advantages of putting a GUI on top of a DOS program. Clearly, there are some definite advantages. Graphics output becomes much more attractive and more flexible, and the equations window of MLwiN is a useful addition, which would be impossible under DOS. There are also some disadvantages, which are the same as for other programs with the same history such as SAS, SPSS and BMDP. We are forced to leave the safe domain of objective and impartial software comparison here, in order to voice an opinion. The GUI of all these programs is ugly as sin. This is partly due to the inherent ugliness of the Windows environment, but also to the fact that the interface is an afterthought. Its only job is to prepare parameter and data files, which are then passed to the DOS programs running in the background. One does not need a GUI to prepare parameter files – in fact, this is not the natural way to prepare them at all. If you are a command line program at heart, then you should acknowledge that, and you should not pretend that a layer of Windows make-up can change this.

A command line interface that builds up the parameter file by asking questions, that is the interface for VARCL and that was also the interface for older versions of HLM has some advantages over the Windows interfaces. It is obviously more structured, and the user does not have to search around in the menus for the appropriate commands. Strange and impossible combinations of options are more easily avoided. It is more foolproof. On the other hand, it rapidly becomes very boring to go through the same sequence of questions every time, and more experienced users will prefer to type the answers ahead of time into a command file and use the program in batch mode (which makes the 'interface' similar to the batch versions of MIXFOO or MLA).

If we summarise our findings, there is little doubt that MLwiN is the most comprehensive program for multilevel analysis. Of course, it is not exactly fair to compare MLwiN with the whole SAS or even the BMDP system. Those are more comprehensive, and they can also be used to perform multilevel analysis. HLM obviously has also developed into a mature and excellent product. Its basic design is more modest than MLwiN, it lacks the macro language, the graphics, and the NANOSTAT statistics and data handling. MLA is somewhat limited, but it can do many useful things, some of which are impossible or very hard to do in the other packages.

Probably for most academic users, the price of the packages is not really a problem. The fact that MLA is free and that MLwiN costs money does not seem to be a major problem. We must say, however, that the suit of MIXFOO programs by Hedeker is really a bargain. One can do ordinary multilevel analysis, multilevel survival analysis, categorical and ordinal outcomes, and Poisson multilevel analysis – and all this in a uniform and modular way. The programs are fast, have little overhead, are versatile, well documented and completely free. Moreover, binaries are available for Power-Macs and for Sun Solaris machines. For most of the MIXFOO programs, one needs to write batch setup files, but, as we have indicated above, that is not really a major disadvantage. Market forces have conspired to make us believe that it is, and that graphical interfaces are the be-all and end-all of software, but most quality statistical and numerical software does not have and does not need such interfaces. Editing an existing text file is usually faster than wading through not very intuitive menus.

References

Aitkin MA, Longford NT (1986) Statistical modeling issues in school effectiveness studies (with discussion). *Journal of the Royal Statistical Society A* **149**, 1–43.

Albert J, Chib S (1997) Bayesian tests and model diagnostics in conditionally independent hierarchical models. *Journal of the American Statistical Association* **92**, 916–25.

Amemiya T (1981) Qualitative response models: A survey. *Journal of Economic Literature* **19**, 483–536.

Belsley DA, Kuh E, Welsch RE (1980) *Regression Diagnostics: Identifying Influential Data and Sources of Collinearity*. Wiley, New York.

Bentham G, Langford IH (1996) Climate change and the incidence of food poisoning in England and Wales. *International Journal of Biometeorology* **39**, 81–86.

Bentham G, Langford IH and Day RJ (2000) Weekly incidence of food poisoning and ambient temperatures in the hot summers of 1989 and 1990. *International Journal of Biometeorology* (submitted).

Bernardinelli L, Montomoli C (1992) Empirical Bayes versus fully Bayesian analysis of geographical variation in disease risk. *Statistics in Medicine* **11**, 983–1007.

Bernardo JM, Smith AFM (1994). B*ayesian Theory*. Wiley, Chichester.

Besag J, York J, Mollié A (1991) Bayesian image restoration, with two applications in spatial statistics. *Annals of the Institute of Statistical Mathematics* **43**, 1–59.

Best NG, Spiegelhalter D, Thomas A, Brayne CEG (1996) Bayesian analysis of realistically complex models. *Journal of the Royal Statistical Society A* **159**, 323–42.

BMDP Statistical Software (1992) BMDP *User's Digest*. University of California Press.

Breslow NE, Clayton DG (1993). Approximate inference in generalized linear mixed models. *Journal of the American Statistical Association* **88**, 9–25.

Bryk AS, Raudenbush SW (1992) *Hierachical Linear Models*. Sage, Newbury Park, CA.

Bryk AS, Raudenbush S, Congdon R (1996) *Hierarchical Linear and Nonlinear Modeling with the HLM/2L and HLM/3L Programs*. Scientific Software International, Chicago.

Bull JM, Riley GD, Rasbash J, Goldstein H (1999) Parallel implementation of a multi-level modelling package. *Computational Statistics and Data Analysis* **31**, 457–74.

Busing FMTA, Meier E, van der Leeden R (1994) MLA software for multilevel analysis of data with two levels. Technical Report, Department of Psychometrics and Research Methodology, University of Leiden.

Cavalli-Sforza LL, Menozzi P, Piazza, A (1994). *The History and Geography of Human Genes*. Princeton University Press.

Chatterjee S, Hadi AS (1988) *Sensitivity Analysis in Linear Regression*. Wiley, New York.

Clayton D (1988) The analysis of event history data: a review of progress and outstanding problems. *Statistics in Medicine* **7**, 819–41.

Clayton, D, Hills, M (1993). *Statistical Models in Epidemiology*. Oxford University Press.

Clayton, D, Kaldor J (1987) Empirical Bayes estimates of age-standardized relative risks for use in disease mapping. *Biometrics* **43**, 671–81.

Clayton D, Rasbash J (1999) Estimation in large crossed random effects models by data augmentation. *Journal of the Royal Statistical Society* A **162**, 425–36.

Cochran WG (1977) *Sampling Techniques*, 3rd edn. Wiley, New York.

Cohen J (1992) A power primer. *Psychological Bulletin* **112**, 155–9.

Cohen M (1998) Determining sample sizes for surveys with data analyzed by hierarchical linear models. *Journal of Official Statistics* **14**, 267–275.

Collett D (1991) *Modelling Binary Data*. Chapman & Hall, London.

Cook RD, Weisberg S (1982, 1995) *Residuals and Influence in Regression*. Chapman & Hall, London.

Copas J (1999) What works? Selectivity models and meta-analysis. *Journal of the Royal Statistical Society A* **162**, 95–109.

Cox DR, Oakes D (1984) *Analysis of Survival Data*. Chapman & Hall, London.

Cox DR, Snell EJ (1989) *Analysis of Binary Data, 2nd edn. Chapman & Hall, London.*

Curtis SL, Diamond I, McDonald JW (1993) Birth interval and family effects on postneonatal mortality in Brazil. *Demography* **30**, 33–43.

Davis P, Gribben B (1995). Rational prescribing and interpractitioner variation: a multilevel approach. *International Journal of Technology Assessment in Health Care* **11**, 428–442.

de Courcy-Wheeler RHB, Wolfe CDA, Spencer M, Goodman JDS, Gamsu HR (1995). Use of the CRIB (clinical risk index for babies) score in prediction of neonatal mortality and morbidity. *Archives of Disease in Childhood* **73**, 32–6.

de Leeuw J, Kreft IGG (1995) Questioning multilevel models. *Journal of Educational and Behavioral Statistics* **20**, 171–90.

DerSimonian R, Laird N (1986) Meta-analysis in clinical trials. *Controlled Clinical Trials* **7**, 177–88.

Diggle PJ (1988) An approach to the analysis of repeated measurements. *Biometrics* **44**, 959–971.

Dixon WJ (1992) BMDP *Statistical Software Manual*. University of California Press.

Dixon WJ, Merdian K (1992) *ANOVA and Regression with* BMDP5–V. Dixon Statistical Associates, Los Angeles.

Dobson AJ (1991) *An Introduction to Generalized Linear Models*. Chapman & Hall, London.

Donner A (1986) A review of inference procedures for the intraclass correlation-coefficient in the one-way random effects model. *International Statistical Review* **54**, 67–82.

DuBois RW, Brook RH, Rogers WH (1987) Adjusted hospital death rates: A potential screen for quality of medical care. *American Journal of Public Health* **77**, 1162–7.

Duncan C (1997) Applying mixed multivariate multilevel models in geographical research. In: Westert GP, Verhoeff RN (eds) *Places and People: Multilevel modelling in Geographical Research*. The Royal Dutch Geographical Society, Utrecht.

Duncan C, Jones K, Moon G (1998) Context, composition and heterogeneity: using multilevel models in health research. *Social Science and Medicine* **46**, 97–117.

Echoard R, Clayton D (1998) Multilevel modelling of conception in artificial insemination by donor. *Statistics in Medicine* **17**, 1137–56.

Elliott P, Martuzzi M, Shaddick G (1995) Spatial statistical methods in environmental epidemiology: a critique. *Statistical Methods in Medical Research* **4**, 137–59.

English D (1992) Geographical epidemiology and ecological studies. In: Elliott P, Cuzick J, English D, Stern R (eds) *Geographical and Environmental Epidemiology: Methods for Small-Area Studies. Oxford University Press, 3–13.*

Epstein A (1995) Performance reports on quality-prototypes, problems and prospects. *New England Journal of Medicine* **333**, 57–61.

Flay BR, Brannon BR, Johnson CA, Hansen WB, Ulene AL, Whitney-Saltiel DA, Gleason LR, Sussman S, Gavin M, Glowacz KM, Sobol DF Spiegel DC (1989) The television, school and family smoking cessation and prevention project: I. theoretical basis and program development. *Preventative Medicine* **17**, 585–607.

Flay BR, Miller TQ, Hedeker D, Siddiqui O, Britton CF, Brannon BR, Johnson CA, Hansen WB, Sussman S, Dent C (1995) The television school and family smoking prevention and cessation project: VIII. Student outcomes and mediating variables. *Preventive Medicine*, **24**, 29–40.

Fuller P, Picciotto A, Davies M, McKenzie SA (1998) Cough and sleep in inner city children. *European Respiratory Journal* **12**, 426–431.

Gatsonis C, Normand S-L, Liu C, Morris C (1993) Geographical variation of procedure utilization: a hierarchical model approach. *Medical Care (Suppl)* **31**, YS55–9.

Gelfand AE and Smith AFM (1990) Sampling-based approaches to calculating marginal densities. *Journal of the American Statistical Society* **85**, 398–409.

Geman S, Geman D (1984) Stochastic relaxation, Gibbs distributions and the Bayesian restoration of images. *IEEE Transactions on Pattern Analysis and Machine Intelligence.* **6**, 721–41.

Gilks WR, Richardson S, Spiegelhalter DJ (1996), *Markov Chain Monte Carlo in Practice.* Chapman & Hall, London.

Goldstein H (1979) *The Design and Analysis of Longitudinal Studies.* Academic Press, London.

Goldstein H (1986) Multilevel mixed linear model analysis using iterative generalized least squares. *Biometrika* **73**, 43–56.

Goldstein H (1987) *Multilevel Models in Educational and Social Research.* Griffin, London.

Goldstein H (1989a) Efficient prediction models for adult height. In: Tanner JM (ed.) *Auxology 88; Perspectives in the Science of Growth and Development.* Smith Gordon, London.

Goldstein H (1989b) Restricted (unbiased) iterative generalized least squares estimation. *Biometrika* **76**, 622–623.

Goldstein H (1991). Nonlinear multilevel models, with an application to discrete response data. *Biometrika* **78**, 45–51.

Goldstein H (1995) *Multilevel Statistical Models, 2nd edn. Edward Arnold, London Wiley, New York.*

Goldstein H, Rasbash J (1992) Efficient computational procedures for the estimation of parameters in multilevel models based on iterative generalized least squares. *Computational Statistics and Data Analysis* **13**, 63–71.

Goldstein H, Rasbash J (1996) Improved approximations for multilevel models with binary responses. *Journal of the Royal Statistical Society A* **159**, 505–13.

Goldstein H, Spiegelhalter DJ (1996) League tables and their limitations: Statistical issues in comparisons of institutional performance. *Journal of the Royal Statistical Society A* **159**, 385–443.

Goldstein H, Rasbash J, Yang M, Woodhouse G, Pan H, Nuttall D (1993) A multilevel analysis of school examination results. *Oxford Review of Education* **19**, 425–33.

Goldstein H, Healy MJR, Rasbash, J (1994) Multilevel time series models with applications to repeated measures data. *Statistics in Medicine* **13**, 1643–55.

Goldstein H, Rasbash J, Plewis I, Draper D, Browne W, Yang M, Woodhouse G, Healy M (1998). *A User's Guide to MLwiN.* University of London.

Goldstein H, Yang M, Omar R, Turner R, Thompson SG (1999) Meta-analysis using multilevel models with an application to the study of class size effects. *Applied Statistics* **49**, 399–412.

Grizzle JC, Allen DM (1969) An analysis of growth and dose response curves. *Biometrics* **25**, 357–61.

Groenewegen P (1997) Dealing with micro-macro relations: a heuristic approach with examples from health services research. In: Westert GP, Verhoeff RN (eds) *Places and People: Multilevel Modelling in Geographical Research.* The Royal Dutch Geographical Society, Utrecht, 9–18.

Guilkey DK, Murphy JL (1993) Estimation and testing in the random effects probit model. *Journal of Econometrics* **59**, 301–17.

Gustafsson JE, Stahl PE (1997) *STREAMS User's Guide.* Multivariate Ware, Mölndal, Sweden.

Harrison GA, Brush G (1990) On correlations between adjacent velocities and accelerations in longitudinal growth data. *Annals of Human Biology* **17**, 55–7.

Hedeker D (1989) Random regression models with autocorrelated errors. PhD Thesis, University of Chicago.

Hedeker D, Gibbons RD (1994) A random effects ordinal regression model for multilevel analysis. *Biometrics* **50**, 933–44.

Hedeker D, Gibbons RD (1996a) MIXOR. A computer program for mixed-effects ordinal probit and logistic regression analysis. *Computer Methods and Programs in Biomedicine* **49**, 157–76.

Hedeker D, Gibbons RD (1996b) MIXREG. A computer program for mixed-effects regression analysis with autocorrelated errors. *Computer Methods and Programs in Biomedicine* **49**, 229–52.

Hedeker D, Gibbons RD, Flay BR (1994) Random-effects regression models for clustered data: with an example from smoking prevention research. *Journal of Consulting and Clinical Psychology* **62**, 757–65.

Hedeker D, Siddiqui O, Hu FB (1996) Random-effects regression analysis of correlated grouped-time survival data. Technical Report, University of Illinois, Chicago.

Hedges L, Vevea JL (1996) Estimating effect size under publication bias: small sample properties and robustness of a random effects selection model. *Journal of Educational and Behavioural Statistics* **21**, 299–332.

Hill PW, Goldstein H (1998) Multilevel modelling of educational data with cross-classification and missing identification of units. *Journal of Educational and Behavioral Statistics* **23**, 117–28.

Hocking RR (1996) *Methods and Applications of Linear Models: Regression and the Analysis of Variance*. Wiley, New York.

Hox JJ (1995) *Applied Multilevel Analysis*. TT-Publikaties, Amsterdam.

Jencks SF, Daley J, Draper D, Thomas N, Lenhart G, Walker J (1988) Interpreting hospital mortality data. The role of clinical risk adjustment. *Journal of the American Medical Association* **260**, 3611–16.

Jennrich R, Schluchter M (1986) Unbalanced repeated measures models with structured covariance matrices. *Biometrics* **42**, 805–20.

Jones K (1997) Multilevel approaches to modelling contextuality: from nuisance to substance in the analysis of voting behaviour. In: Westert GP, Verhoeff RN (eds) *Places and People: Multilevel Modelling in Geographical Research*. The Royal Dutch Geographical Society, Utrecht 19–43.

Jones K, Moon G, Clegg, A (1991) Ecological and individual effects in childhood immunisation uptake: a multi-level approach. *Social Science and Medicine* **33**, 501–8.

Kenward MG (1998) Selection models for repeated measurements with non-random dropout: an illustration of sensitivity. *Statistics in Medicine* **17**, 2723–32.

Knaus WA, Draper EA, Wagner DP, Zimmerman JE (1985) APACHE II: A Severity of Disease Classification System. *Critical Care Medicine* **13**, 818–29.

Kraemer HC, Thiemann S (1987) *How many subjects? Statistical power analysis in research*. Sage, London.

Kreft IGG, de Leeuw J (1998) Introducing Multilevel Modelling. Sage, London.

Kreft IGG, de Leeuw J, Kim K-S (1990) Comparing four different statistical packages for hierarchical linear regression: GENMOD, HLM, ML2, VARCL. UCLA Statistics Preprint 50.

Kreft IGG, de Leeuw J, van der Leeden R (1994) Review of five multilevel analysis programs: BMDP-5V, GENMOD, HLM, ML3, VARCL. *American Statistician* **48**, 324–35.

Kunin CM, Johansen KS, Worning AM, Daschner FD (1990) Report of a symposium on use and abuse of antibiotics worldwide. *Reviews of Infectious Diseases* **12**, 12–19.

Lange N, Ryan L (1989) Assessing normality in random effects models. *Annals of Statistics* **17**, 624–42.

Langford IH (1994) Using empirical Bayes estimates in the geographical analysis of disease risk. *Area* **26.2**, 142–9.

Langford IH, Bentham G (1997) A multi-level model of sudden infant death syndrome in England and Wales. *Environment and Planning A* **29**, 629–40.

Langford I, Lewis, T (1998) Outliers in multilevel data. *Journal of the Royal Statistical Society A* **161**, 121–60.

Langford IH, Bentham G, McDonald A-L (1998) Multilevel modelling of geographically aggregated health data: a case study on malignant melanoma mortality and UV exposure in the European Community: a multi-level modelling approach. *Statistics in Medicine* **17**, 41–58.

Langford IH, Leyland AH, Rasbash J, Goldstein H (1999a) Multilevel modelling of the geographical distributions of rare diseases. *Applied Statistics* **48**, 253–68.

Langford IH, Leyland AH, Rasbash J, Goldstein H, Day RJ, McDonald A-L (1999b) Multilevel modelling of area-based health data. In: Lawson A, Biggeri A, Böhning D, Lesaffre E, Viel J-F, Bertolini R (eds) *Disease Mapping and Risk Assessment for Public Health*. Wiley, Chichester, 217–28.

Lawson AB, Biggeri A, Williams FLR. (1999a) A review of modelling approaches in health risk assessment around putative sources. In: Lawson A, Biggeri A, Böhning D, Lesaffre E, Viel J-F, Bertolini R (eds) *Disease Mapping and Risk Assessment for Public Health*. Wiley, Chichester, 231–45.

Lawson AB, Böhning D, Biggeri A, Lesaffre E, Viel J-F. (1999b) Disease mapping and its uses. In: Lawson A, Biggeri A, Böhning D, Lesaffre E, Viel J-F, Bertolini R (eds) *Disease Mapping and Risk Assessment for Public Health*. Wiley, Chichester, 3–13.

Leyland A H, Boddy F A (1998) League tables and acute myocardial infarction. *Lancet* **351**, 555–8.

Leyland AH, Langford IH, Rasbash J, Goldstein H (1998) Spatio-temporal modelling of mortality data for multiple causes of death. In: *Proceedings of the Biometrics Section of the American Statistical Association*, 10–15.

Leyland AH, Langford IH, Rasbash J, Goldstein H (2000) Multivariate spatial models for event data. *Statistics in Medicine* **19**, 2469–2478.

Li XS (1997) multilevel statistical modelling for discrete responses in epidemiological field study. PhD Dissertation, School of Public Health, West China University of Medical Sciences.

Lindsey JK (1999) Relationships among sample size, model selection and likelihood regions, and scientifically important differences. *Journal of the Royal Statistical Society D* **48**, 401–12.

Littell RC, Milliken GA, Stroup WW, Wolfinger RD (1996) *SAS System for Mixed Models*. SAS Institute, Cary, NC.

Little RJA (1995). Modeling the dropout mechanism in repeated measures studies. *Journal of the American Statistical Association* **90**, 1112–21.

Longford NT (1988) A quasi-likelihood adaptation for variance component analysis. In: *Proceedings of the Section on Statistical Computing of the ASA*, 137–42.

Longford NT (1990) VARCL Software for variance component analysis of data with nested random effects (maximum likelihood). Technical Report, Educational Testing Service, Princeton.

Longford NT (1993) *Random Coefficient Models*. Oxford University Press.

McCullagh P, Nelder JA (1994). *Generalized Linear Models*. Chapman & Hall, London.

McCulloch C, Leyland AH, Boddy FA (1997) Understanding the relationship between length of stay and readmission rates for hospitals (progress report for the King's Fund Comparative Database Initiative). University of Glasgow, Public Health Research Unit.

McDonald RP (1994) The bilevel reticular action model for path analysis with latent variables. *Sociological Methods and Research* **22**, 399–413.

McKee M, Hunter D (1995). Mortality league tables: Do they inform or mislead? *Quality in Health Care* **4**, 5–12.

McLeod A, Ross P, Mitchell S, Tay D, Hunter L, Hall A, Paton J, Mutch L (1996) Respiratory health in a total very low birthweight cohort and their classroom controls. *Archives of Disease in Childhood* **74**, 188–94.

Maddala GS (1996). *Limited Dependent and Qualitative Variables in Econometrics*. Cambridge University Press.

Marshall EC Spiegelhalter DJ (1998) League tables of in vitro fertilisation clinics: How confident can we be about the rankings? *British Medical Journal* **317**, 1701–04

Modigliani F, Ando A (1963) The "life-cycle" hypothesis of saving: aggregate implications and tests. *American Economic Review* **53**, 55–84.

Moerbeek M, van Breukelen GJP, Berger MPF (2000) Design issues for experiments in multilevel populations. *Journal of Educational and Behavioural Statistics* **25**, 271–284.

Moerbeek M, van Breukelen GJP, Berger MPF (2001) Optimal experimental designs for multilevel logistic models. *The Statistician* **50**, 17–30.

Mok M (1995) Sample size requirements for 2–level designs in educational research. *Multilevel Modelling Newsletter* **7(2)**, 11–15 [available from http://www.ioe.ac.uk/multilevel/workpap.html].

Mollié A (1996) Bayesian mapping of disease. In: Gilks WR, Richardson S, Spiegelhalter DJ (eds). *Markov Chain Monte Carlo In Practice*. Chapman & Hall, Boca Raton FL, 360–79.

Mollié A (1999) Bayesian and empirical Bayes approaches to disease mapping. In: Lawson A, Biggeri A, Böhning D, Lesaffre E, Viel J-F, Bertolini R (eds) *Disease Mapping and Risk Assessment for Public Health*. Wiley, Chichester 15–29.

Morris CN, Christiansen CL (1996). Hierarchical models for ranking and for identifying extremes, with applications. In: Bernardo JO, Berger JO, Dawid, AP, AFM (eds) *Bayesian Statistics 5*. Oxford University Press, 277–97.

Muthén BO (1994) Multilevel covariance structure analysis. *Sociological Methods and Research* **22**, 376–98.

Muthén LK, Muthén BO (1998) *Mplus User's Guide*. Muthén & Muthén, Los Angeles.

New York State Department of Health (1995) *Coronary Artery Bypass Surgery in New York State, 1991–1993, New York State Department of Health, Albany*.

New York State Department of Health (1996) *Coronary Arteny Bypass Surgery in New York State, 1992–1994*, New York State Department of Health, Albary.

NHS Executive (1995) *The NHS Performance Guide 1994–1995*. NHS Executive, Leeds.

Normand S-L T, Glickman ME, Ryan TJ (1995). Modelling mortality rates for elderly heart attack patients: profiling hospitals in the co-operative cardiovascular project. In: Gatsonis C, Hodges J, Kass RN, Singpurwalla N (eds) Case Studies in Bayesian Statistics. Springer-Verlag, New York, 435–56.

O'Donnell O, Propper C, Upward, R. (1993). United Kindom. In: Van Doorslaer E, Wagstaff A, Rutten F (eds) *Equity in the Finance and Delivery of Health Care: An International Perspective* Oxford University Press, 237–261.

OPCS (Office of Population Censuses and Surveys) (1989, 1990) *Communicable Disease Statistics: Series MB12*. HMSO, London.

Palmer MJ, Phillips BF, Smith GT (1991) Application of nonlinear models with random coefficients to growth data. *Biometrics* **47**, 623–35.

Pan HQ, Goldstein H (1998) Multilevel repeated measures growth modelling using extended spline functions. *Statistics in Medicine* **17**, 2755–70.

Pfefferman D, Skinner CJ, Holmes DJ, Goldstein H, Rasbash J (1998) Weighting for unequal selection probabilities in multilevel models. *Journal of the Royal Statistical Society B* **60**, 23–40.

Plewis I (1985). *Analysing Change*. Wiley, Chichester.

Plewis I (1993) Reading progress. In: Woodhouse G (ed.) *A Guide to ML3 for New Users* Multilevel Models Project, Institute of Education, London.

Pocock SJ (1983) *Clinical Trials: A Practical Approach*. Wiley, Chichester.

Ramsey JB (1969) Tests for specification errors in classical linear least squares regression analysis. *Journal of the Royal Statistical Society B* **31**, 350–71.

Rasbash J, Goldstein H (1994) Efficient analysis of mixed hierarchical and cross-classified random structures using a multilevel model. *Journal of Educational and Behavioral Statistics* **19**, 337–50.

Rasbash J, Browne W, Goldstein H, Yang M et al. (1999a) *A user's guide to MLwiN*, 2nd edn. Institute of Education, London.

Rasbash J, Healy M, Browne W, Cameron B, Charlton C (1999b) *MLwiN v1.01*. Multilevel Models Project, Institute of Education, London.

Rasbash J, Woodhouse G (1995) *MLn Command Reference*. Institute of Education, London.

Raudenbush SW (1997) Statistical analysis and optimal design for cluster randomized trials. *Psychological Methods* **2**, 173–185.

Rice N, Leyland A (1996) Multilevel models: applications to health data. *Journal of Health Services Research and Policy* **1**, 154–64.

Richardson S, Green PJ (1997) On Bayesian analysis of mixtures with an unknown number of components. *Journal of the Royal Statistical Society B* **59**, 731–92.

Ripley B (1987) *Stochastic Simulation*. Wiley, New York.

Rodriguez G, Goldman N (1995) An assessment of estimation procedure for multilevel models with binary responses. *Journal of the Royal Statistical Society A* **158**, 73–89.

Rothman KJ, Boice JD (1979) *Epidemiological Analysis with a Programmable Calculator*. NIH Publication 79–1649. US Government Printing Office, Washington, DC.

Rowan KM, Kerr JH, McPherson K, Short A, Vessey MP (1993) Intensive care society's APACHE II study in Britain and Ireland – II: Outcome comparisons of intensive care units after adjustment for case mix by the American APACHE II method. *British Medical Journal* **307**, 977–81.

Rubin DB (1976) Inference and missing data. *Biometrika* **63**, 581–92.

Rubin DB (1987) *Multiple Imputation for Nonresponse in Surveys*. Wiley, New York.

SAS (1992) SAS/STAT software: changes and enhancements. Technical report, SAS Institute, Cary, NC.

Schluchter M (1988) BMDP5–V – unbalanced repeated measures models with structured covariance matrices. Technical Report 86, BMDP Statistical Software, Los Angeles.

Schneider EC, Epstein AM (1996) Influence of cardiac-surgery performance reports on referral practices and access to care. *New England Journal of Medicine* **335**, 251–6.

Scott A, Shiell A (1997a) Do fee descriptors influence treatment choices in general practice? A multilevel discrete choice model. *Journal of Health Economics* **16**, 323–42.

Scott A, Shiell A (1997b) Analysing the effect of competition on general practitioners' behaviour using a multilevel modelling framework. *Health Economics* **6**, 577–88.

Scottish Office (1995) *Clinical Outcome Indicators – 1994*. Clinical Resource and Audit Group, Edinburgh.

Singer JD (1998) Using SAS PROC MIXED to fit multilevel models, hierarchical models, and individual growth models. *Journal of Educational and Behavioral Statistics* **23**, 323–55.

Skinner CJ, Holt D, Smith TMF (eds) (1989) *Analysis of Complex Surveys*. Wiley, New York.

Smans M, Muir CS, Boyle P (1992) *Atlas of Cancer Mortality in the European Economic Community*, International Agency for Research or Cancer, Lyon.

Snijders TAB, Bosker RJ (1993) Standard errors and sample sizes for two-level research. *Journal of Educational Statistics* **18**, 237–59.

Snijders TAB, Bosker RJ (1999) *Multilevel Analysis. An Introduction to Basic and Advanced Multilevel Modeling. Sage, London.*

Spiegelhalter DJ (1998) Bayesian graphical modelling: a case study in monitoring health outcomes. *Applied Statistics* **47**, 115–34.

Spiegelhalter DJ, Thomas A, Best NG, Gilks WR (1995) BUGS: Bayesian inference using Gibbs sampling, version 0.50. Technical Report, MRC Biostatistics Unit, Cambridge.

Steele F, Diamond I, Amin S (1996) Immunisation uptake in rural Bangladesh: a multilevel analysis. *Journal of the Royal Statistical Society A* **159**, 289–99.

Steen Carlsson K, Lyttkens CH (1997) *Health Care Consumption and Individual Consumption Possibilities*. Department of Community Medicine, University of Malmö.

Stern H, Cressie N (1999) Inference for extremes in disease mapping. In: Lawson A, Biggeri A, Böhning D, Lesaffre E, Viel J-F, Bertolini R (eds) *Disease Mapping and Risk Assessment for Public Health*. Wiley, Chichester, 63–84.

Taylor A (1994) Appendix: Sample characteristics, attrition and weighting. In: Buck N, Gershuny J, Rose D, Scott J (eds) *Changing Households: The British Household Panel Survey 1990–92*. ESRC Research Centre on Micro-Social Change, colchester, Essex.

Thomas N, Longford NT, Rolph JE (1994) Empirical Bayes methods for estimating hospital specific mortality rates. *Statistics in Medicine* **13**, 889–903.

Thompson, SG, Pocock SJ (1991) Can meta-analyses be trusted? *Lancet* **338**, 1127–30.

van der Leeden R, Vrijburg K, de Leeuw J (1991) A review of two different approaches for the analysis of growth data using longitudinal mixed linear models: Comparing hierarchical linear regression (ML3, HLM) and repeated measures design with structured covariance matrices (BMDP-5V). UCLA Statistics Preprint 98.

van der Leeden R, Vrijburg K, de Leeuw J (1996) A review of two different approaches for the analysis of growth data using longitudinal mixed linear models. *Computational Statistics and Data Analysis* **21**, 583–605.

Van Doorslaer E, Wagstaff A, Rutten, F (eds) (1993) *Equity in the Finance and Delivery of Health Care: An International Perspective*. Oxford University Press.

Wagstaff A, Van Doorslaer, E. (1993) Equity in the finance and delivery of health care: concepts and definitions. In: Equity in the Finance and Delivery of Health Care: An international perspective. Van Doorslaer E, Wagstaff A, Rutten F (eds). Oxford University Press, 7–19.

Wakefield J, Elliott P (1999) Issues in the statistical analysis of small area health data. *Statistics in Medicine* **18**, 2377–99.

WHO (1991a) *Technical bases for the WHO Recommendations on the Management of Pneumonia in Children at First-Level Health Facilities*. WHO/ARI/91.20, Geneva.

WHO (1991b) *Programme for Control of ARI. Case Management of the Young Child with an ARI*. WHO, Geneva.

Woodhouse G (1995) *A Guide to MLn for New Users*. Multilevel Models Project, Institute of Education, London.

Woodhouse G, Yang M, Goldstein H, Rasbash J (1996) Adjusting for measurement error in multilevel analysis. *Journal of the Royal Statistical Society A* **159**, 201–12.

Wright DB (1997) Extra-binomial variation in multilevel logistic models with sparse structures. *British Journal of Mathematical and Statistical Psychology* **50**, 21–9.

Yang M (1997) Multilevel models for multiple category responses – a simulation, *Multilevel Modelling Newsletter*, *Multilevel Models Project, Institute of Education*, Vol. 9, No. 1, 9–15.

Yang M, Goldstein H, Heath A (2000) Multilevel models for repeated binary outcomes: attitudes and voting over the electoral cycle. *Journal of the Royal Statistical Society Series A* **163**, 49–62.

Yang M, Rasbash J, Goldstein H (1998) *MLwiN Macros for Advanced Multilevel Modelling*. Institute of Education, London.

Yang M, Rasbash J, Goldstein H, Barbosa M (1999) *MLwiN Macros for Advanced Multilevel Modelling*. Institute of Education, London.

Index

WILEY SERIES IN PROBABILITY AND STATISTICS
ESTABLISHED BY WALTER A. SHEWHART AND SAMUEL S. WILKS

Editors
Peter Bloomfield, Noel A. C. Cressie, Nicholas I. Fisher, Iain M. Johnstone, J. B. Kadane, Louise M. Ryan, David W. Scott, Bernard W. Silverman, Adrian F. M. Smith, Jozef L. Teugels
Editors Emeritus
Vic Barnett, Ralph A. Bradley, J. Stuart Hunter, David G. Kendall

Probability and Statistics Section

---

*Now available in a lower priced paperback edition in the Wiley Classics Library.

*Now available in a lower priced paperback edition in the Wiley Classics Library.

---

*Now available in a lower priced paperback edition in the Wiley Classics Library.

*Now available in a lower priced paperback edition in the Wiley Classics Library.

*Now available in a lower priced paperback edition in the Wiley Classics Library.

---

*Now available in a lower priced paperback edition in the Wiley Classics Library.

Texts and References Section

AGRESTI · An Introduction to Categorical Data Analysis
ANDERSON · An Introduction to Multivariate Statistical Analysis, *Second Edition*
ANDERSON and LOYNES · The Teaching of Practical Statistics
ARMITAGE and COLTON · Encyclopedia of Biostatistics. 6 Volume set
BARTOSZYNSKI and NIEWIADOMSKA-BUGAJ · Probability and Statistical
    Inference
BENDAT and PIERSOL · Random Data: Analysis and Measurement Procedures, *Third
    Edition*
BERRY, CHALONER, and GEWEKE · Bayesian Analysis in Statistics and
    Econometrics: Essays in Honor of Arnold Zellner
BHATTACHARYA and JOHNSON · Statistical Concepts and Methods
BILLINGSLEY · Probability and Measure, *Second Edition*
BOX · R. A. Fisher, the Life of a Scientist
BOX, HUNTER, and HUNTER · Statistics for Experimenters: An Introduction to
    Design, Data Analysis, and Model Building
BOX and LUCEÑO · Statistical Control by Monitoring and Feedback Adjustment
BROWN and HOLLANDER · Statistics: A Biomedical Introduction
CHATTERJEE and PRICE · Regression Analysis by Example, *Third Edition*
COOK and WEISBERG · An Introduction to Regression Graphics
COOK and WEISBERG · Applied Regression Including Computing and Graphics
COX · A Handbook of Introductory Statistical Methods
DILLON and GOLDSTEIN · Multivariate Analysis: Methods and Applications
*DODGE and ROMIG · Sampling Inspection Tables, *Second Edition*
DRAPER and SMITH · Applied Regression Analysis, *Third Edition*
DUDEWICZ and MISHRA · Modern Mathematical Statistics
DUNN · Basic Statistics: A Primer for the Biomedical Sciences, *Second Edition*
EVANS, HASTINGS and PEACOCK · Statistical Distributions, *Third Edition*
FISHER and VAN BELLE · Biostatistics: A Methodology for the Health Sciences
FREEMAN and SMITH · Aspects of Uncertainty: A Tribute to D. V. Lindley
GROSS and HARRIS · Fundamentals of Queueing Theory, *Third Edition*
HALD · A History of Probability and Statistics and their Applications Before 1750
HALD · A History of Mathematical Statistics from 1750 to 1930
HELLER · MACSYMA for Statisticians
HOEL · Introduction to Mathematical Statistics, *Fifth Edition*
HOLLANDER and WOLFE · Nonparametric Statistical Methods, *Second Edition*
HOSMER and LEMESHOW · Applied Logistic Recession, *Second Edition*
HOSMER and LEMESHOW · Applied Survival Analysis: Regression Modeling of Time to
    Event Data
JOHNSON and BALAKRISHNAN · Advances in the Theory and Practice of Statistics:
    A Volume in Honor of Samuel Kotz
JOHNSON and KOTZ (editors) · Leading Personalities in Statistical Sciences: From the
    Seventeenth Century to the Present
JUDGE, GRIFFITHS, HILL, LÜTKEPOHL, and LEE · The Theory and Practice of
    Econometrics, *Second Edition*
KHURI · Advanced Calculus with Applications in Statistics
KOTZ and JOHNSON (editors) · Encyclopedia of Statistical Sciences. Volumes 1 to 9
    with Index
KOTZ and JOHNSON (editors) · Encyclopedia of Statistical Sciences: Supplement
    Volume

*Now available in a lower priced paperback edition in the Wiley Classics Library.

WILEY SERIES IN PROBABILITY AND STATISTICS

ESTABLISHED BY WALTER A. SHEWHART AND SAMUEL S. WILKS

Editors
Robert M. Groves, Graham Kalton, J. N. K. Rao, Norbert Schwarz, Christopher Skinner

Survey Methodology Section

*Now available in a lower priced paperback edition in the Wiley Classics Library.

COCHRAN · Sampling Techniques, *Third Edition*

COUPER, BAKER, BETHLEHEM, CLARK, MARTIN, NICHOLLS and O'REILLY (editors) · Computer Assisted Survey Information Collection

COX, BINDER, CHINNAPPA, CHRISTIANSON, COLLEDGE, and KOTT (editors) · Business Survey Methods

*DEMING · Sample Design in Business Research

DILLMAN · Mail and Telephone Surveys: The Total Design Method

GROVES · Survey Errors and Survey Costs

GROVES and COUPER · Nonresponse in Household Interview Surveys

GROVES, BIEMER, LYBERG, MASSEY, NICHOLLS, and WAKSBERG · Telephone Survey Methodology

*HANSEN, HURWITZ, and MADOW · Sample Survey Methods and Theory, Volume I: Methods and Applications

*HANSEN, HURWITZ, and MADOW · Sample Survey Methods and Theory, Volume II: Theory

KISH · Statistical Design for Research

*KISH · Survey Sampling

KORN and GRAUBARD · Analysis of Health Surveys

LESSLER and KALSBEEK · Nonsampling Error in Surveys

LEVY and LEMESHOW · Sampling of Populations: Methods and Applications

LYBERG, BIEMER, COLLINS, de LEEUW, DIPPO, SCHWARZ, TREWIN (editors) · Survey Measurement and Process Quality

SIRKEN, HERRMANN, SCHECHTER, SCHWARZ, TANUR and TOURNANGEAU (editors) · Cognition and Survey Research

VALLIANT, DORFMAN, and ROYALL · Finite Population Sampling and Inference: A Prediction Approach

---

*Now available in a lower priced paperback edition in the Wiley Classics Library.